AF183303

Metallurgy: Principles and Processes

Metallurgy: Principles and Processes

Chloe Raines

Willford Press
www.willfordpress.com

Published by Willford Press,
118-35 Queens Blvd., Suite 400,
Forest Hills, NY 11375, USA

Copyright © 2019 Willford Press

This book contains information obtained from authentic and highly regarded sources. All chapters are published with permission under the Creative Commons Attribution Share Alike License or equivalent. A wide variety of references are listed. Permissions and sources are indicated; for detailed attributions, please refer to the permissions page. Reasonable efforts have been made to publish reliable data and information, but the authors, editors and publisher cannot assume any responsibility for the validity of all materials or the consequences of their use.

Trademark Notice: Registered trademark of products or corporate names are used only for explanation and identification without intent to infringe.

ISBN: 978-1-68285-732-8

Cataloging-in-Publication Data

Metallurgy : principles and processes / Chloe Raines.
p. cm.
Includes bibliographical references and index.
ISBN 978-1-68285-732-8
1. Metallurgy. 2. Materials science. I. Raines, Chloe.
TN665 .M48 2019
669--dc23

For information on all Willford Press publications
visit our website at www.willfordpress.com

WILLFORD PRESS

Printed in China.

Contents

Preface

Metallurgy is the process of separation of metals from their ores. It is also an academic field which analyzes the properties of metallic elements, inter-metallic compounds and alloys. The extraction of metals requires the processing of ores through various processes. Accordingly, this field can be categorized into physical and chemical metallurgy. Depending on the materials produced, metallurgy branches into ferrous and non-ferrous metallurgy. Various techniques of plating, metalworking, thermal spraying, shot peening and heat treatment are crucial in metallurgy. This textbook provides comprehensive insights into the principles and processes in the field of metallurgy. It presents this complex subject in the most comprehensible and easy to understand language. This book is an essential guide for both academicians and those who wish to pursue this discipline further.

A detailed account of the significant topics covered in this book is provided below:

Chapter 1- The domain of metallurgy is concerned with the study of the physical and chemical behavior of metals or metallic elements, their separation from ores, intermetallic compounds and alloys. This chapter has been carefully written to provide an introduction to metallurgy and its varied processes.

Chapter 2- The primary subdivisions of metallurgy are ferrous and non-ferrous metallurgy. Nearly 95 percent of the world metal production involves ferrous metals. The aim of this chapter is to explore the common types of metallurgy, such as extractive metallurgy, physical metallurgy, chemical metallurgy, electrometallurgy, etc.

Chapter 3- Metal extraction involves the practice of removing metals from an ore through physical, chemical and electrolytical processes and then refining them into a purer state. The topics elaborated in this chapter on the concentration of ore, isolation of metal from concentrated ore, purification of metal, etc. will help in providing a better perspective about metal extraction.

Chapter 4- A combination of metals or metal and an element is called an alloy. The metals such as aluminium, iron, copper, zinc, chromium, etc. are often used for making alloys. The diverse topics elaborated in this chapter such as substitutional alloys, interstitial alloys, amalgam, amorphous metal, oxide dispersion strengthened (ODS) alloy, shape-memory alloy, etc. will provide a comprehensive understanding of the subject matter.

Chapter 5- In order to completely understand metallurgy, it is necessary to understand the processes related to it, such as Bayer process, Bessemer process, Betts electrolytic process, patio process, Pidgeon process, Hoopes process, etc. which have been thoroughly discussed in this chapter.

It gives me an immense pleasure to thank our entire team for their efforts. Finally in the end, I would like to thank my family and colleagues who have been a great source of inspiration and support.

Chloe Raines

1

Introduction to Metallurgy

The domain of metallurgy is concerned with the study of the physical and chemical behavior of metals or metallic elements, their separation from ores, intermetallic compounds and alloys. This chapter has been carefully written to provide an introduction to metallurgy and its varied processes.

Metallurgy is defined as a process that is used for the extraction of metals in their pure form. The compounds of metals mixed with soil, limestone, sand, and rocks are known as minerals. Metals are commercially extracted from minerals at low cost and minimum effort. These minerals are known as ores. A substance which is added to the charge in the furnace to remove the gangue (impurities) is known as flux. Metallurgy deals with the process of purification of metals and the formation of alloys.

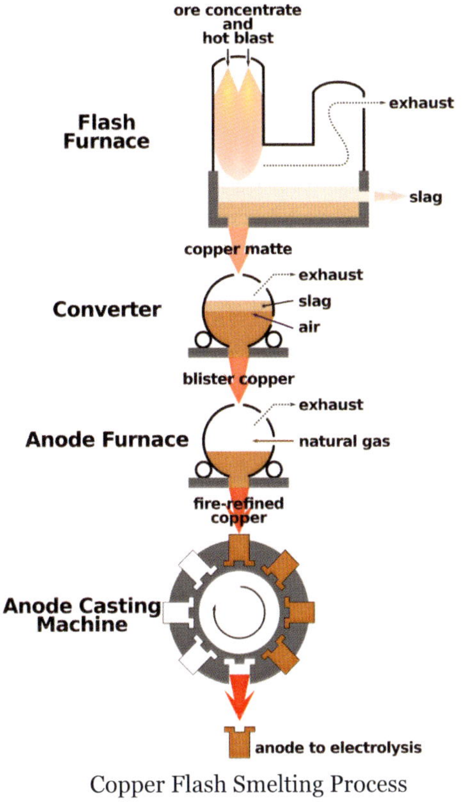

Copper Flash Smelting Process

Principles of Metallurgy

The metallurgical process can be classified as the following:

1. Crushing and grinding: The first process in metallurgy is crushing of ores into a fine powder in a crusher or ball mill. This process is known as pulverization.

2. The concentration of ores: The process of removing impurities from ore is known as a concentration of minerals or ore dressing. In metallurgy, we concentrate the ores mainly by the following methods.

3. Hydrolytic method: In this method, we pour the ore over a sloping, vibrating corrugated table with grooves. A jet of water is allowed to flow over the surface. The denser ore particles settle in the grooves, and the impurities are washed away by water.

4. Magnetic separation: In this case, the crushed ore is placed on a conveyor belt. This belt rotates around two wheels in which one of the wheels is magnetic, and therefore the magnetic particles get attracted to the magnetic wheel and fall apart from the non-magnetic particles.

5. Froth floatation: In this process, we take the crushed ore in a large tank which contains oil and water. A current of compressed air is passed through it. The ore gets wet by oil and is separated from the impurities in the form of froth. Ore is lighter, and so it comes on the surface and impurities are left behind.

6. Roasting and calcination: In metallurgy, the process of heating a concentrated ore in the presence of oxygen is known as roasting. This process is applied in a case of sulfide ores. For ores containing carbonate or hydrated oxides, heating is done in the absence of air to melt the ores, and this process is known as calcination.

Types of Metallurgy

Extractive Metallurgy

Extractive metallurgy is the practice of removing valuable metals from an ore and refining the extracted raw metals into a purer form. In order to convert a metal oxide or sulfide to a purer metal, the ore must be reduced physically, chemically, or electrolytically. Extractive metallurgists are interested in three primary streams: feed, concentrate (valuable metal oxide/sulfide), and tailings (waste).

After mining, large pieces of the ore feed are broken through crushing and/or grinding. This step creates particles that are either mostly valuable or mostly waste. Concentrating the particles of value in a form supporting separation enables the desired metal to be removed from waste products.

Ore bodies often contain more than one valuable metal. Tailings of a previous process may be used as a feed in another process to extract a secondary product from the original ore. Additionally, a concentrate may contain more than one valuable metal. That concentrate would be processed to separate the valuable metals into individual constituents.

Hydrometallurgy

Hydrometallurgy is concerned with processes that use aqueous solutions to extract metals from ores. The most common hydrometallurgical process is leaching, which involves dissolution of the valuable metals into the aqueous solution. After the solution is separated from the ore solids, the solution is often subjected to various processes of purification and concentration before the valuable metal is recovered, either in its metallic state or as a chemical compound. The solution purification and concentration processes may include precipitation, distillation, adsorption, and solvent extraction. The final recovery step may involve precipitation, cementation, or an electrometallurgical process.

Sometimes, hydrometallurgical processes may be carried out directly on the ore material without any pre-treatment steps. More often, the ore must be pre-treated by various mineral processing steps and sometimes by pyrometallurgical processes.

Pyrometallurgy

Pyrometallurgy involves high temperature processes where chemical reactions take place among gases, solids, and molten materials. Solids containing valuable metals are either reacted to form intermediate compounds for further processing, or they are converted into their elemental or metallic state. Pyrometallurgical processes that involve gases and solids are typified by roasting operations. Processes that produce molten products are collectively referred to as smelting operations. The energy required to sustain the high temperature pyrometallurgical processes may come entirely from the exothermic nature of the chemical reactions taking place, usually oxidation reactions. Often, however, energy must be added to the process by combustion of fuel or, in the case of some smelting processes, by the direct application of electrical energy.

Electrometallurgy

Electrometallurgy involves metallurgical processes that take place in some form of electrolytic cell. The most common types of electrometallurgical processes are electrowinning and electro-refining. Electrowinning is an electrolysis process used to recover metals in aqueous solution, usually as the result of an ore having undergone one or more hydrometallurgical processes. The metal of interest is plated onto a cathode, while an anode is composed of an inert electrical conductor. Electro-refining is used to dissolve an impure metallic anode (typically from a smelting process) and produce a high purity cathode. Fused salt electrolysis is another electrometallurgical process whereby the valuable metal is dissolved into a molten salt, which acts as the electrolyte, and the valuable metal collects on the cathode of the cell.

The scope of electrometallurgy has significant overlap with the areas of hydrometallurgy and (in the case of fused salt electrolysis) pyrometallurgy. Additionally, electrochemical phenomena play a considerable role in many mineral processing and hydrometallurgical processes.

Metallurgy in Production Engineering

In production engineering, metallurgy is concerned with the production of metallic components for use in consumer or engineering products. This involves the production of alloys, the shaping, the heat treatment and the surface treatment of the product. The task of the metallurgist is to achieve design criteria specified by the mechanical engineer, such as cost, weight, strength, toughness, hardness, corrosion and fatigue resistance, and performance in temperature extremes.

Common engineering metals are aluminum, chromium, copper, iron, magnesium, nickel, titanium, and zinc. These are most often used as alloys. Much effort has been placed on understanding one very important alloy system, that of purified iron, which has carbon dissolved in it, better known as steel. Normal steel is used in low cost, high strength applications where weight and corrosion are not a problem. Cast irons, including ductile iron are also part of this system.

Stainless steel or galvanized steel are used where resistance to corrosion is important. Aluminium alloys and magnesium alloys are used for applications where strength and lightness are required.

Most engineering metals are stronger than most plastics and are tougher than most ceramics. Composites of plastics and materials such as glass fiber and carbon fiber rival metals in applications requiring high tensile strength with little weight. Concrete rivals metals in applications requiring high compressive strength and resistance to the effects of water. Wood rivals metal in applications requiring low cost and availability of materials and low cost of construction, as well as in applications requiring certain aesthetics.

The operating environment of the product is very important—a well-designed material will resist expected failure modes such as corrosion, stress concentration, metal fatigue, creep, and environmental stress fracture. Ferrous metals and some aluminium alloys in water and especially in an electrolytic solution such as seawater, corrode quickly. Metals in cold or cryogenic conditions tend to lose their toughness becoming more brittle and prone to cracking. Metals under continual cyclic loading can suffer from metal fatigue. Metals under constant stress in hot conditions can creep.

Production Engineering of Metals

A blacksmith works at his forge.

Metals are shaped by processes such as casting, forging, rolling, extrusion, sintering, metalworking, machining and fabrication. With casting, molten metal is poured into a shaped mold. With forging, a red-hot billet is hammered into shape. With rolling, a billet is passed through successively narrower rollers to create a sheet. With extrusion, a hot and malleable metal is forced under pressure through a die, which shapes it before it cools. With sintering, a powdered metal is compressed into a die at high temperature. With machining, lathes, milling machines, planning machines and drills are used to cut the cold metal to shape. With fabrication, sheets of metal are cut with guillotines or gas cutters and bent into shape.

“Cold working” processes, such as rolling and fabrication, where the product’s shape is altered while the product is cold, can increase the strength of the product by a process called work hardening. Work hardening creates microscopic defects in the metal, which resist further changes of shape.

Various forms of casting exist in industry and academia. These include sand casting, investment casting (also called the "lost wax process"), die casting, and continuous casting.

Welding is a technique for joining certain ferrous metals and certain aluminum alloys. The metals in the weld and on both sides of the join are generally similar alloys. Brazing is a technique for joining copper-based metals.

Metals can be heat treated by annealing, quenching, tempering, and case hardening to alter properties of toughness, hardness or resistance to corrosion. Annealing is used to make a shaped product tougher by reducing the effects of work hardening, it also softens the metal. Quenching and case hardening are used to make a shaped product harder. Quenching by it self makes the metal very hard and very brittle, tempering after quenching is used to reduce the brittleness and improve overall properties.

Electroplating is the main surface treatment technique and involves bonding a thin layer of another protective metal such as gold, silver, chromium, or zinc to the surface of the product to reduce corrosion.

Electrical and Electronic Engineering

Metallurgy is also applied to electrical and electronic materials where metals such as aluminum, copper, tin, and gold are used in power lines, wires, printed circuit boards, and integrated circuits.

Soldering is a method of joining metallic electrical conductors where high strength is not required.

Metallurgical Techniques

Metallurgists study the microscopic and macroscopic mechanisms that cause a metal or alloy to behave in the way that it does—that is, the changes that occur on the atomic level that affect the metal's (or alloy's) macroscopic properties. Examples of tools used for microscopic examination of metals are optical and electron microscopes and mass spectrometers.

Metallurgists study crystallography, the effects of temperature and heat treatment on the component phases of alloys, such as the eutectic, and the properties of those alloy phases.

The macroscopic properties of metals are tested using machines and devices that measure tensile strength, compressive strength, and hardness.

2

Common Types of Metallurgy

The primary subdivisions of metallurgy are ferrous and non-ferrous metallurgy. Nearly 95 percent of the world metal production involves ferrous metals. The aim of this chapter is to explore the common types of metallurgy, such as extractive metallurgy, physical metallurgy, chemical metallurgy, electrometallurgy, etc.

Extractive Metallurgy

Extractive metallurgy is the practice of removing valuable metals from an ore and refining the extracted raw metals into a purer form. In order to convert a metal oxide or sulfide to a purer metal, the ore must be reduced physically, chemically, or electrolytically. Extractive metallurgists are interested in three primary streams: feed, concentrate (valuable metal oxide/sulfide), and tailings (waste).

After mining, large pieces of the ore feed are broken through crushing and/or grinding. This step creates particles that are either mostly valuable or mostly waste. Concentrating the particles of value in a form supporting separation enables the desired metal to be removed from waste products.

Ore bodies often contain more than one valuable metal. Tailings of a previous process may be used as a feed in another process to extract a secondary product from the original ore. Additionally, a concentrate may contain more than one valuable metal. That concentrate would be processed to separate the valuable metals into individual constituents.

Concentration

After mining, large pieces of the ore feed are broken through crushing and/or grinding. This step creates particles that are either mostly valuable or mostly waste. Concentrating the particles of value in a form supporting separation enables the desired metal to be removed from waste products. Froth flotation is one process for separating minerals from the surrounding worthless material by taking advantage of differences in their *hydrophobicity*. whereby hydrophobicity differences between valuable minerals other material are increased through the use of surfactants and wetting agents. Flotation is used for the separation of a large range of sulfides, carbonates and oxides prior to further refinement. Phosphates and coal are also upgraded (purified) by flotation technology.

Other ways to concentration minerals include approached based on density (Froth Flotation) used if ore density is less than density of impurities, or via melting points, or magnetic properties (Magnetic separation). This can be used with magnetic ores (i.e. Fe_3O_4) that are passed through a magnetic field, which attracts the magnetic ore and ignore the nonmagnetic impurities.

A cylindrical froth flotation cell.

Roasting

Ore is crushed and heated to a high temp using a strong blast of hot air. The process converts the ores to their oxides which can then be reduced. Consider the natural occurring ores of zinc (ZnS (sphalerite) and $ZnCO_3$ (smithsonite). When roasted, smithsonite decomposes to $ZnO_{(s)}$ and $CO_2(g)$ $ZnCO_{3(s)} \xrightarrow{\Delta} ZnO_{(s)} + CO_{2(g)}$ and the hot air involved in roasting sphalerite oxidizes it $2ZnS_{(s)} + 3O_{2(g)} \xrightarrow{\Delta} 2ZnO_{(s)} + 2SO_{2(g)}$ to produces $ZnO_{(s)}$ and $SO_2(g)$.

$$ZnCO_{3(s)} \xrightarrow{\Delta} ZnO_{(s)} + CO_{2(g)}$$

$$2ZnS_{(s)} + 3O_{2(g)} \xrightarrow{\Delta} 2ZnO_{(s)} + 2SO_{2(g)}$$

Reduction

$C_{(s)}$ and $CO_{(g)}$ are often used as reducing agents in simultaneous reactions. Oxides of Cr, V, and Mn are reduced using Al. During the reduction process the metal oxide is heated to a temperature above its boiling point in order to vaporize it and condense as a liquid. For example, $ZnO_{(s)}$ that is produced in the roasting process is combined with $C_{(s)}/CO_{(g)}$ mixture and heated at a 1100 °C. Two reduction reactions are viable at this temperature with either carbon ($ZnO_{(s)} + C_{(s)} \xrightarrow{\Delta} Zn_{(g)} + CO_{(g)}$) or carbon monoxide ($ZnO_{(s)} + CO_{(g)} \xrightarrow{\Delta} Zn_{(g)} + CO_{2(g)}$) acting the reducing agent.

$$ZnO_{(s)} + C_{(s)} \xrightarrow{\Delta} Zn_{(g)} + CO_{(g)}$$

$$ZnO_{(s)} + CO_{(g)} \xrightarrow{\Delta} Zn_{(g)} + CO_{2(g)}$$

Refining

The impurities contained in the metal product of the roasting/reduction process are removed. For example, the Zn(l) produced in the reduction process often contains impurities of Cd and Pb. Though fractional distillation of Zn(l) will work in the refining process, a more commonly used and efficient method is electrolysis. The $ZnO_{(s)}$ produced from roasting is dissolved in $H_2SO_{4(aq)}$ and

$Zn_{(s)}$ powder is added allowing the impurities to be displaced. The solution is then electrolyzed and to get a pure metallic $Zn_{(s)}$ as a result.

Zone Refining

A rod containing the desired pure metal and impurities is passed through a series of heating coils and cooled again. The process isolates the impurities that concentrate in the molten zones, leaving the portions behind them somewhat more pure. This process is repeated until impurities are moved to the end of the rod and cut off, resulting in an almost completely pure metal rod.

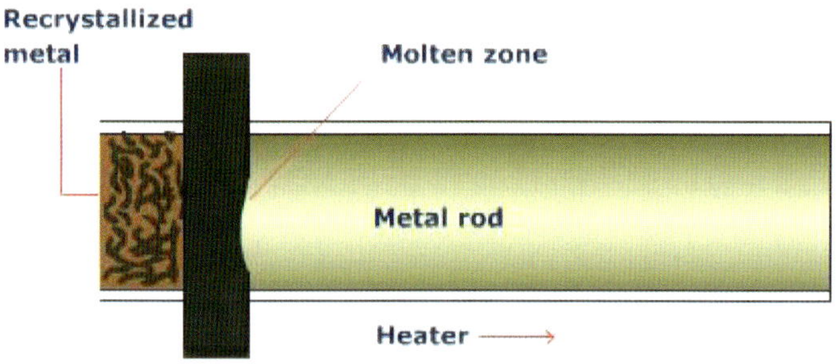

The Zone refining method.

Pyrometallurgy

Two of the most common pyrometallurgical processes, in both extraction and refining, are oxidation and reduction. In oxidation, metals having a great affinity for oxygen selectively combine with it to form metallic oxides; these can be treated further in order to obtain a pure metal or can be separated and discarded as a waste product. Reduction can be viewed as the reverse of oxidation. In this process, a metallic oxide compound is fed into a furnace along with a reducing agent such as carbon. The metal releases its combined oxygen, which recombines with the carbon to form a new carbonaceous oxide and leaves the metal in an uncombined form.

Oxidation and reduction reactions are either exothermic (energy-releasing) or endothermic (energy-absorbing). One example of an exothermic reaction is the oxidation of iron sulfide (FeS) to form iron oxide (FeO) and sulfur dioxide (SO_2) gas:

$$2FeS + 3O_2 \rightarrow 2FeO + 2SO_2$$

This process gives off large quantities of heat beyond that required to initiate the reaction. One endothermic reaction is the smelting reduction of zinc oxide (ZnO) by carbon monoxide (CO) to yield zinc (Zn) metal and carbon dioxide (CO_2):

$$ZnO + CO \rightarrow Zn + CO_2$$

For this reaction to proceed at a reasonable rate, external heat must be supplied to maintain the temperature at 1,300 to 1,350 °C (2,375 to 2,450 °F).

Roasting

For those instances in which a metal-bearing compound is not in a chemical form that permits the metal to be easily and economically removed, it is necessary first to change it into some other compound. The preliminary treatment that is commonly used to do this is roasting.

Processes

There are several different types of roast, each one intended to produce a specific reaction and to yield a roasted product (or calcine) suitable for the particular processing operation to follow. The roasting procedures are:

1. Oxidizing roasts, which remove all or part of the sulfur from sulfide metal compounds, replacing the sulfides with oxides. (The sulfur removed goes off as sulfur dioxide gas.) Oxidizing roasts are exothermic.
2. Sulfatizing roasts, which convert certain metals from sulfides to sulfates. Sulfatizing roasts are exothermic.
3. Reducing roasts, which lower the oxide state or even completely reduce an oxide to a metal. Reducing roasts are exothermic.
4. Chloridizing roasts, or chlorination, which change metallic oxides to chlorides by heating with a chlorine source such as chlorine gas, hydrochloric acid gas, ammonium chloride, or sodium chloride. These reactions are exothermic.
5. Volatilizing roasts, which eliminate easily volatilized oxides by converting them to gases.
6. Calcination, in which solid material is heated to drive off either carbon dioxide or chemically combined water. Calcination is an endothermic reaction.

Roasters

Each of the above processes can be carried out in specialized roasters. The types most commonly in use are fluidized-bed, multiple-hearth, flash, chlorinator, rotary kiln, and sintering machine (or blast roaster).

Fluidized-bed roasters have found wide acceptance because of their high capacity and efficiency. They can be used for oxidizing, sulfatizing, and volatilizing roasts. The roaster is a refractory-lined, upright cylindrical steel shell with a grate bottom through which air is blown in sufficient volume to keep fine, solid feed particles in suspension and give excellent gas-solid contact. The ore feed can be introduced dry or as a water suspension through a downpipe into the turbulent layer zone of the roaster. Discharge of the roasted calcines is through a side overflow pipe.

Multiple-hearth roasters also have found wide acceptance in that they can be used for oxidizing, sulfatizing, chloridizing, volatilizing, reducing, and calcining processes. The roaster is a refractory-lined, vertical cylindrical steel shell in which are placed a number of superimposed refractory hearths. A slowly rotating central shaft turns rabble arms on each hearth both to stir the roasting material and to push it into drop holes leading to the hearth below. Feed material is fed to the top hearth, and, as it follows a zig-zag path across the hearths and downward, it meets the rising gas stream that effects the roasting. The calcines are discharged from the bottom hearth.

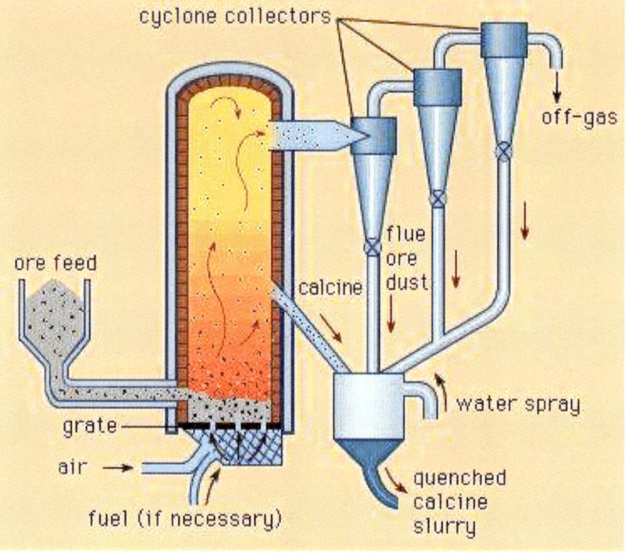

A fluidized-bed roaster

Flash roasters are used only for oxidizing roasts and are, in effect, multiple-hearth roasters with the central hearths removed. This design came with the realization that much of the oxidizing takes place as the particles are actually dropping from hearth to hearth.

Chlorinators are used for roasting oxides to chlorides. They are tall, circular steel shells lined with refractory brick to prevent chlorine attack on the steel. The top of each chlorinator has a sealed hopper for periodic feed charging, and gaseous or liquid chlorine is added at the bottom of the unit. Heat is supplied by electrical resistance through the shell wall and by any exothermic reaction that may occur. The product depends on the chloridizing reaction taking place, with magnesium dichloride, for example, forming as a watery liquid and titanium tetrachloride coming off as a gas.

Calcination of carbonates to oxides is done in a horizontal rotary kiln, which is a mild-steel circular shell lined with refractory material and having a length 10 to 12 times the diameter. Sloping slightly downward from feed to discharge ends, the kiln slowly rotates while fuel-fired burners located inside the kiln provide the required heat.

A sintering machine, or blast roaster, can conduct oxidizing or reducing roasts and then agglomerate the roasted calcines, or it can be used for agglomeration alone. (Agglomeration is the fusing of fine feed material into larger chunks that can be fed into a blast furnace or retort, eliminating the problem of losing the fine feed in the hot air blast.) The oxidizing or reducing reaction is exothermic, but in order for agglomeration alone to be conducted, a fuel such as fine coke must be mixed with the charge.

The sintering machine consists of an endless belt of moving metal pallets with grate bottoms on which a fine feed charge is spread and passed under a burner. As the charge ignites, the pallet passes over a suction wind box, so that air being drawn through the feed layer causes combustion (*i.e.,* oxidation) of sulfur or carbon to continue from top to bottom. Because the temperature is high and there is no agitation of the feed, a partial fusion takes place on the surface of the particles, leaving them adhering together in the form of a porous, cellular clinker known as sinter.

Smelting

Smelting is a process that liberates the metallic element from its compound as an impure molten metal and separates it from the waste rock part of the charge, which becomes a molten slag. There are two types of smelting, reduction smelting and matte smelting. In reduction smelting, both the metallic charge fed into the smelter and the slag formed from the process are oxides; in matte smelting, the slag is an oxide while the metallic charge is a combination of metallic sulfides that melt and recombine to give a homogeneous metallic sulfide called matte.

Reduction Smelting

Many types of furnace are used for reduction smelting. The blast furnace is universally used in the reduction of such compounds as iron oxide, zinc oxide, and lead oxide, though there are great differences between the furnace designs used in each case. Iron, found naturally in the oxide ores hematite and magnetite, is smelted in a tall, circular, sealed blast furnace. A sintered or pelletized feed consisting of coke (for fuel), limestone (as a flux for slag making), and iron oxide is charged into the top of the furnace through a double bell or rotating chute, and heated air is blown in through nozzles, or tuyeres, close to the furnace bottom. In the ensuing combustion reaction, oxygen in the air combines with carbon in the coke, generating enough heat to melt the furnace charge and forming carbon monoxide, which, in turn, reduces the iron oxide to metallic iron. The furnace is sealed to prevent the escape of carbon monoxide gas, which is recovered and burned as fuel to heat the tuyere air. In the hearth at the furnace bottom, molten slag and iron collect in two layers, the lighter slag on top. Both are periodically drawn off, with the slag being discarded and the iron going on to be refined into steel.

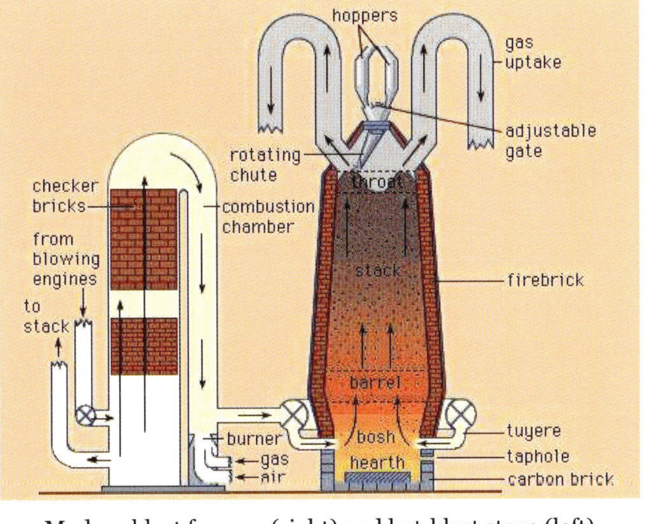

Modern blast furnace (right) and hot-blast stove (left)

The zinc blast furnace also is a sealed furnace, with a charge of sintered zinc oxide and preheated coke added through a sealed charging bell. The furnace is rectangular, with a shorter shaft than the iron blast furnace. A blast of hot air through the tuyeres provides oxygen to burn the coke for heat and to supply carbon monoxide reducing gas. The reduced zinc passes out of the furnace as vapour, and this is drawn off to a spray bath of molten lead and condensed to liquid zinc metal. Slag and any lead present in the charge are tapped as liquids from the furnace hearth.

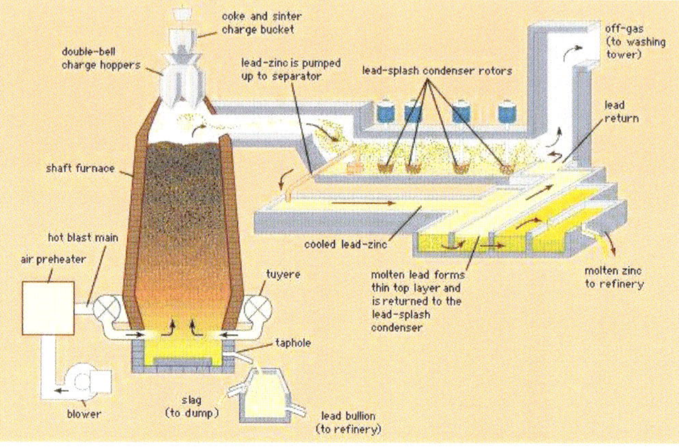

A zinc-lead blast furnace and lead-splash condenser.

The lead blast furnace is similar in size and shape to the zinc blast furnace, but it is not a sealed furnace, and it does not use preheated tuyere air. A charge of lead oxide sinter, coke, and flux is poured into the open top of the furnace, and the strong reducing atmosphere in the furnace shaft reduces the oxide to metal. Liquid lead and slag collect in two layers in the furnace hearth, with the lead in the bottom layer and slag above.

Two newer processes for the direct reduction of unroasted lead sulfide concentrate are the QSL (Queneau-Schuhmann-Lurgi) and the KIVCET (a Russian acronym for "flash-cyclone-oxygen-electric smelting"). In the QSL reactor a submerged injection of shielded oxygen oxidizes lead sulfide to lead metal, while the KIVCET is a type of flash-smelting furnace in which fine, dried lead sulfide concentrate combines with oxygen in a shaft to give lead metal.

Matte Smelting

The primary purpose of matte smelting is to melt and recombine the charge into a homogeneous matte of metallic copper, nickel, cobalt, and iron sulfides and to give an iron and silicon oxide slag. It is done in many types of furnace on both roasted or unroasted sulfide feed material.

The reverberatory furnace is essentially a rectangular refractory-brick box equipped with end-wall burners to provide heat for melting. The furnace is relatively quiet, and it does not blow out much fine feed (which is added through roof ports) with the exhaust gases. The matte is tapped periodically from a centre taphole, while the slag runs off continuously at the furnace flue end. Oxygen lances inserted through the roof, or oxygen added through the burners, can increase smelting capacity considerably.

Electric furnaces are similar to reverberatory furnaces except for the method of heating—in this case a row of electrodes projecting through the roof into the slag layer on the furnace hearth and heating by resistance.

Flash smelting is a relatively recent development that has found worldwide acceptance. It is an autogenous process, using the oxidation of sulfides in an unroasted charge to supply the heat required to reach reaction temperatures and melt the feed material. The most widely used furnace has a vertical reaction shaft at one end of a long, low settling hearth and a vertical gas-uptake shaft at the other end. Fine, unroasted feed is blown into the reaction shaft along with preheated air;

these react instantaneously, and liquid droplets fall onto the settling hearth, separating into layers of slag and matte. The off-gas, high in sulfur dioxide, is ideal for sulfur-recovery processes.

The second stage of matte smelting is converting the sulfides to metal. For many years the standard vessel for this operation has been the Peirce-Smith converter. This is a rotatable, refractory-lined, horizontal steel drum with an opening at the centre of the top for charging and discharging and a row of tuyeres across the back through which air, oxygen-enriched air, or oxygen can be blown into the liquid bath. Molten matte from the smelting furnace is poured into the converter, after which gas is blown through the tuyeres to oxidize first iron and then sulfur. The sulfur goes off as sulfur dioxide gas and the iron as iron oxide slag, leaving semipure metal. Considerable heat is generated by this exothermic reaction, keeping the bath liquid and maintaining the required reaction temperature.

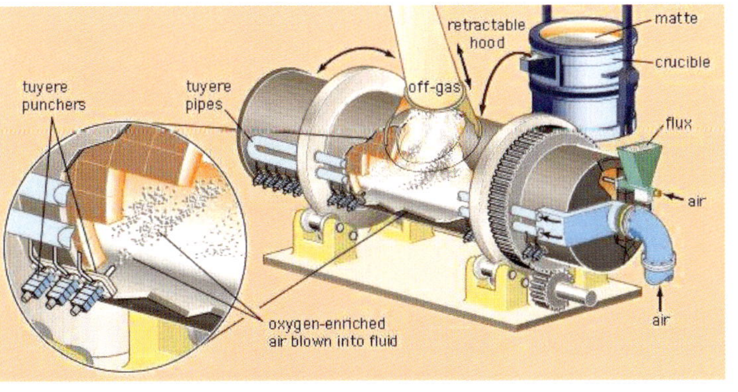

A side-blown copper-nickel matte converter.

More recent processes take advantage of exothermic heat evolution to accomplish both the smelting of unroasted sulfides and the conversion of matte in one combined operation. These are the Noranda, TBRC (top-blown rotary converter), and Mitsubishi processes. The Noranda reactor is a horizontal cylindrical furnace with a depression in the centre where the metal collects and a raised hearth at one end where the slag is run off. Pelletized unroasted sulfide concentrate is poured into the molten bath at one end, where tuyeres inject an air-oxygen mixture. This causes an intense mixing action that aids the melting, smelting, and oxidation steps, which follow one another in sequence, by taking advantage of the exothermic heat. The TBRC also is cylindrical in shape but is inclined at 17° to the horizontal, has an open mouth at the high end for charging and pouring, and revolves at 5 to 40 rotations per minute. A lance inserted through the mouth can give any combination of oxygen, air, or natural gas to impinge on the molten bath and create the conditions required for smelting and oxidizing. The combination of surface blowing and bath rotation improves the performance of the converter. The Mitsubishi process is a continuous smelting-converting operation that uses three stationary furnaces in series. The first furnace is for smelting, with oxygen lances and a fuel-fired burner inserted through the roof. Slag and matte flow from here to a slag-cleaning furnace (heated by electric arc), and high-grade matte flows from this to the converting furnace, where oxygen-enriched air is blown into the bath through roof lances. Exothermic heat produced here is sufficient to keep the bath up to reaction temperature.

Electrolytic Smelting

Smelting is also carried out by the electrolytic dissociation, at high temperatures, of a liquid metallic chloride compound (as is done with magnesium) or of a metallic oxide powder dissolved in molten electrolyte (as is done with aluminum). In each case, electric current is passed through the bath to dissociate the metallic compound; the metal released collects at the cathode, while a gas is given off at the anode.

The magnesium smelting cell consists of a steel pot that serves as the cathode; two rows of graphite electrodes are inserted through a refractory cover as anodes. The electrolyte is a mixture of chlorides, with magnesium chloride making up 20 percent, and the cell is maintained at 700 °C (1,300 °F). The passage of current breaks down the magnesium chloride into chlorine gas and magnesium metal, which go to the anode and cathode, respectively.

In the Hall-Héroult smelting process, a nearly pure aluminum oxide compound called alumina is dissolved at 950 °C (1,750 °F) in a molten electrolyte composed of aluminum, sodium, and fluorine; this is electrolyzed to give aluminum metal at the cathode and oxygen gas at the anode. The smelting cell is a carbon-lined steel box, which acts as the cathode, and a row of graphite electrodes inserted into the bath serves as anodes.

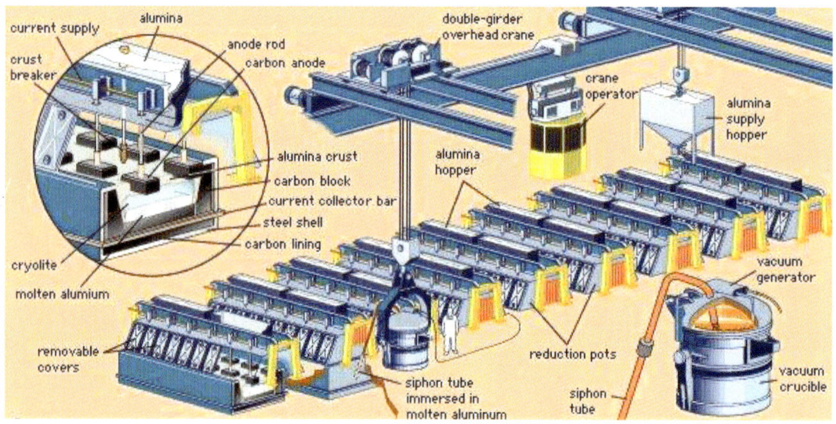

Part of a modern potline based on the electrolytic Hall-Héroult smelting process.

Refining

Refining is the final procedure for removing (and often recovering as by-products) the last small amounts of impurities left after the major extraction steps have been completed. It leaves the major metallic element in a practically pure state for commercial application. The procedure is accomplished in three ways: refining by fire, by electrolytic, or by chemical methods.

Fire Refining

Iron, copper, and lead are fire-refined by selective oxidation. In this process, oxygen or air is added to the impure liquid metal; the impurities oxidize before the metal and are removed as an oxide slag or a volatile oxide gas.

The basic oxygen furnace (BOF) is a vessel used to convert pig iron, of about 94 percent iron and 6 percent combined impurities such as carbon, manganese, and silicon, into steel with as little as 1 percent combined impurities. The BOF is a large pear-shaped unit that can be tilted to charge and pour. Molten blast-furnace iron and steel scrap are charged into the furnace; then it is turned to an upright position and a lance inserted to blow high-tonnage oxygen gas into the bath. Oxidation reactions occur rapidly, with silicon and manganese oxidizing first and combining to form an oxide slag, then carbon oxidizing to carbon monoxide gas and burning to carbon dioxide as it leaves the furnace mouth. These reactions are strongly exothermic and keep the vessel up to its reaction temperature without any external heat or fuel being added.

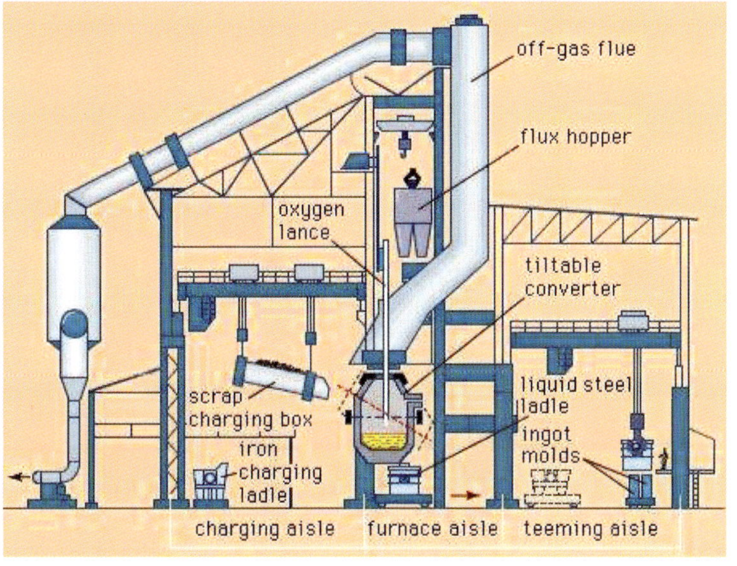

A basic oxygen furnace shop.

Converter-produced blister copper and blast-furnace lead also are treated by fire refining, with both processes depending on the weaker affinity for oxygen of the metals than the impurities they contain. Molten copper in a small reverberatory-type furnace has compressed air blown into it through steel pipes below the surface. This oxidizes zinc, tin, iron, lead, arsenic, antimony, and sulfur; the sulfur goes off as sulfur dioxide gas, while the other impurities form an oxide slag that is skimmed off. Lead is refined in much the same way, with compressed air blown into a molten lead bath and the major impurities of tin, antimony, and arsenic oxidizing in that order, rising to the surface as skims and being scraped off.

Other fire-refining operations use fractional distillation. By this method, zinc metal of 98 percent purity can be upgraded to 99.995 percent purity. The main impurities in blast-furnace zinc are lead and cadmium, with lead boiling at 1,744° C (3,171° F), zinc at 907° C (1,665° F), and cadmium at 765° C (1,409° F). In the first stage zinc and cadmium are boiled off, leaving liquid lead, and in the second stage cadmium is boiled off to leave special high-purity zinc metal.

Electrolytic Refining

This method gives the highest-purity metal product as well as the best recovery of valuable impurities. It is used for copper, nickel, lead, gold, and silver. The metal to be refined is cast into a slab, which becomes the anode of an electrolytic cell; another sheet of metal is the cathode. Both electrodes are immersed in an aqueous electrolyte capable of conducting an electric current. As a direct current is impressed on the cell, metal ions dissolve from the anode and deposit at the cathode. The insoluble sludge left in the cell is treated to recover any valuable by-product metals.

Chemical Refining

An example of chemical refining is the nickel carbonyl process, in which impure nickel metal is selectively reacted with carbon monoxide gas to form nickel carbonyl gas. This gas is then decomposed to give high-purity nickel metal.

Hydrometallurgy

Hydrometallurgy is a process for recovering valuable metals dissolved in a water-based chemical mix containing metallic salt compounds. Such metal recovery usually is based upon sulfuric acid solutions and the metals that are targeted for recovery are either precious metals such as gold, platinum, and palladium, or more common industrial metals like copper, nickel, and zinc. The chemical processes are typically carried out in a titanium reactor that, though susceptible to corrosion by sulfuric acid, is protected by the interaction of the dissolved metals in the solution itself. The reactors are run at high temperatures and pressures, and can employ various methods of removing the metals, though leaching is the most common.

While hydrometallurgy has been practiced as far back as the 16th century, only in recent years has it become widespread due to the increasing value of precious metals like gold. This has made it economically viable to extract such metals from very low concentrations in ore. It is estimated that, as of the year 2011, over 70 different metallic elements were being extracted through hydrometallurgy processes, including such widely varying metals as uranium, mercury, and cobalt.

Another term for hydrometallurgy is solution mining, and its often an important secondary process in industries that regularly use acids such as sulfuric acid and hydrochloric acid. The process has a dual benefit. It both purifies the acids, which are often used as metal cleaning or electrolyte agents, and it separates out the trace metals for resale.

Hydrometallurgy revolves around three processes: leaching, metal recovery and solution purification.

The various techniques used in these processes include:

- Leaching: Autoclave, column, atmospheric and pressure acid depending on the metal ore.
- Recovery: Carbon-in-leach and carbon-in-pulp, oxidation—biological or pressure method—and cementation.
- Solution purification: Ion-exchange, solvent extraction, electrowinning, precipitation and process simulations to determine flow sheet design.

Leaching

Leaching involves the use of aqueous solutions, which is brought into contact with a material containing a valuable metal; the solution may be acidic or basic. In the leaching process, oxidation potential, temperature, and pH of the solution are important parameters, and are often manipulated to optimize dissolution of the desired metal component into the aqueous phase.

Heap Leaching of Uranium

Heap leaching is an extraction process by which chemicals (usually sulfuric acid) are used to extract the economic element from ore which has been mined and placed in piles on the surface. Heap leaching is generally economically feasible only for oxide ore deposits. Oxidation of sulfide deposits occurs during the geological process called weathering. Therefore, oxide ore deposits are typically found close to the surface.

Most uranium is found in the upper crust, occurring naturally as U^{4+} and U^{6+} with an average abundance of 2 ppm. Its most naturally occurring mineral is Uraninite (UO2), which typically reacts with oxygen to form U3O8, the mineral pitchblende. If there are no other economic elements within the ore a mine might choose to extract the uranium using a leaching agent, usually a low molar sulfuric acid. Similar to copper oxide heap leaching, also using dilute sulfuric acid. The final product is yellowcake (U_3O_8) and requires significant further processing to produce fuel-grade feed.

Acid leaching is sometimes also referred to as heap leaching because the leaching process can be performed on large "heaps" of uranium ore that have been collected from mines. The chemistry of the leaching process revolves around oxidation of the uranium compounds, which is typically achieved using manganese dioxide (MnO_2), sodium chlorate ($NaClO_3$), and Fe(II) salts. Equations below shows a typical reaction pathway for the acid leaching process using sulfuric acid.

$$UO_{2(s)} + 2Fe^{3+} \rightleftharpoons UO^{2+}_{2(aq)} + 2Fe^{2+}$$

$$UO_3 + H^+ \rightleftharpoons UO^{2+}_{2(aq)} + H_2O$$

$$UO^{2+}_{2(aq)} + 3SO^{2-}_4 \rightleftharpoons \left[UO_2(SO_4)_3\right]^{4-}_{(aq)}$$

Sulfuric acid is typically used due to the solubility of uranyl sulfate complexes. The reaction is typically performed at slightly elevated temperatures (~60 °C) and can often release H_2, H_2S, and CO_2 gases during the process. The uranium, which typically begins in the tetravalent state, goes through a series of reactions, eventually leading to the formation of the desired complex, $\left[UO_2(SO_4)_3\right]^{4-}_{(aq)}$. While the solubility of this complex makes sulfuric acid a desirable leaching agent, nitric and hydrochloric acid can also be used, but are typically not due to their higher cost and corrosiveness.

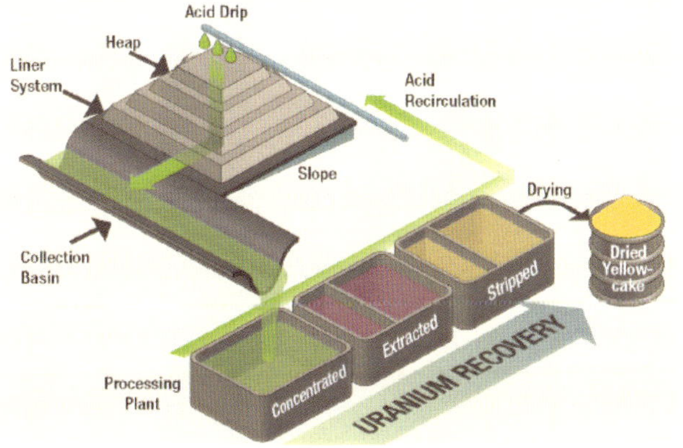

Heap leach recovery for uranium (US NRC).

To convert yellowcake into weapons grade quality, it is converted into uranium hexafluoride (UF_6) ore to be isotopically enriched. UF_6 can be easily turned into a gas for enrichment based off of the differing effusion rates of the different isotopes.

Solution Concentration and Purification

After leaching, the leach liquor must normally undergo concentration of the metal ions that are to be recovered. Additionally, undesirable metal ions sometimes require removal. Precipitation is the selective removal of a compound of the targeted metal or removal of a major impurity by precipitation of one of its compounds. Froth flotation is a process for selectively separating hydrophobic materials from hydrophilic materials.

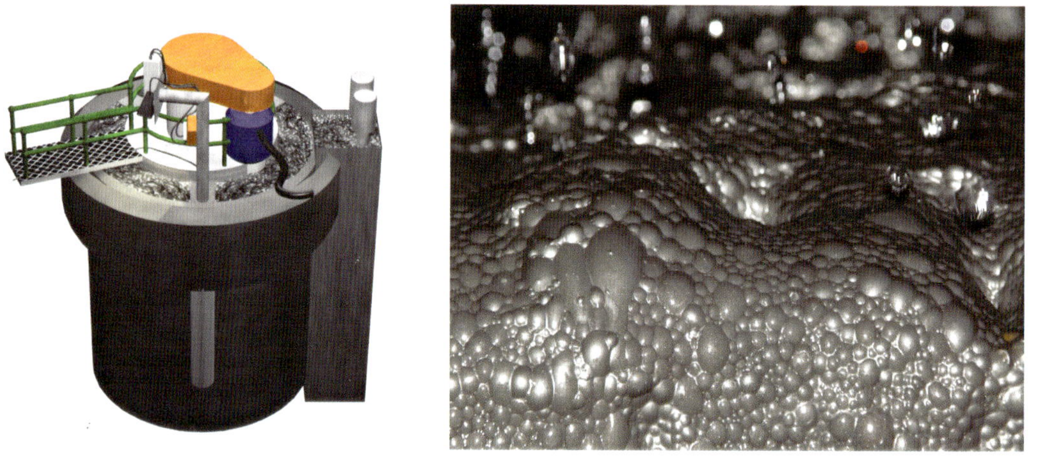

(Left) A cylindrical froth flotation cell (right) Copper-sulfide loaded air bubbles in a froth

Metal Recovery

Metal recovery is the final step in a hydrometallurgical process. Metals suitable for sale as raw materials are often directly produced in the metal recovery step. Sometimes, however, further refining is required if ultra-high purity metals are to be produced. The primary types of metal recovery processes are electrolysis, gaseous reduction, and precipitation. For example, a major target of hydrometallurgy is copper, which is conveniently obtained by electrolysis.

Example

The most selective methods for separating metals from their ores are based on the formation of metal complexes. For example, gold is often found as tiny flakes of the metal, usually in association with quartz or pyrite deposits. In those circumstances, gold is typically extracted by using cyanide leaching, which forms a stable gold–cyanide complex—$[Au(CN)_2]^-$:

$$4Au_{(s)} + 8NaCN_{(aq)} + O_{2(g)} + 2H_2O_{(l)} \rightarrow 4Na\left[Au(CN)_2\right]_{(aq)} + 4NaOH_{(aq)}$$

Virtually pure gold can be obtained by adding powdered zinc to the solution:

$$Zn_{(s)} + 2\left[Au(CN)_2\right]^-_{(aq)} \rightarrow \left[Zn(CN)_4\right]^{2-}_{(aq)} + 2Au_{(s)}$$

A related method, which is used to separate Co^{3+}, Ni^{2+}, and Cu^+ from Fe, Mn, and Ti, is based on the formation of stable, soluble ammonia complexes of ions of the late transition metals.

Non-ferrous Metallurgy

The metals other than iron and alloys that do not contain an appreciable amount of ferrous (iron) are known as non-ferrous metals. A distinguishing feature of non-ferrous metals is that they are highly malleable (i.e., they can be pressed or hammered into thin sheets without breaking).

Non-ferrous metals have one valuable advantage over ferrous metals, which is that they are highly corrosion and rust resistant because they do not have any iron content in them. Consequently, these materials are suitable for highly corrosive environments such as liquid, chemical and sewage pipelines. Non-ferrous metals are also non-magnetic, which make them suitable for many electrical and electronic applications.

Some commonly used non-ferrous metals are copper, zinc, aluminum, lead, nickel, cobalt, chromium, gold, silver and many others.

Some common non-ferrous metals used in industrial processes are:

- Aluminum
- Copper
- Lead
- Zinc
- Tin

General Principles of Extraction of Metals from Oxides & Sulphides

In an extraction process, a specific amount of energy required to reduce or release a metal ion present in a compound (solid or liquid) in association with other elements and to collect the metal in its elemental form.

In nature, metals generally occur either in free form or in combined form. Generally, low reactive metals shows little affinity towards air, moisture, carbon dioxide or other non-metals present in nature. Such metals may remain in elemental or native (free) state in nature. Such metals are called *Noble Metals* as they show the least chemical reactivity. For example gold (Au), silver (Ag), mercury (Hg), and platinum (Pt) etc. Whereas, most of the metals are active to combine with air, moisture, carbon dioxide and non-metals like oxygen, sulphur, halogens etc. to form their compounds, like oxides, sulphides, carbonates, halides and silicates. i.e., in nature, they occur in a combined state.

A naturally occurring material in which a metal or its compound occurs is called a *mineral*. A mineral from which a metal can be extracted economically is called an *ore*. An ore is that mineral in which a metal is present in appreciable quantities and from which the metal can be extracted economically. The main active substances present in nature, expecially in the atmosphere are oxygen and carbon dioxide. In the earth's crust, sulphur (S) and silicon (Si) are found in large quantities. Most active metals are highly electropositive and therefore exist as ions. It is for this reason that most of the important *ores* of these metals occur as

1. Oxides
2. Sulphides
3. Carbonates
4. Halides
5. Silicates

Some sulphide ores undergo oxidation by air to form sulphates. This explains the occurrence of sulphate ores. Ores are invariably found in nature in contact with rocky materials. These rocky or earthy impurities accompanying the ores are termed as *gangue* or matrix.

During metal extraction, the bulk of the metal is separated from the impurities present in the ore by using a system which invariably comprises more than one phase. The metal enters the metallic phase while the impurities enter the non-metallic phase or phases. The driving force for the reactions involved in metal extraction depends on the differences in the chemical potentials of the chemical elements. During a chemical reaction, an element tends to move from a high to a low chemical potential state. In each extraction process, the basic aim is to create conditions conductive to such a transfer, which facilitates the extraction of a metal or its compound.

In metal extraction, a vast majority of metallurgical reactions are made to take place at elevated temperatures because then the ore compounds become relatively unstable, facilitating the release of the metal. This is the basis of pyrometallurgy, in electrometallurgy, the dissociation or decomposition of a compound in solution (i. e. reduction of an iron) is brought about to employing electric forces. In hydrometallurgy, the reduction process depends on the judicious manipulation of the chemical reaction taking place in an aqueous solution.

The process of extracting the metals from their ores and refining them is called *metallurgy*. The choice of the process depends upon the nature of the ore and the type of the metal. The metal content in the ore can vary depending upon the impurities present and chemical composition of the ore. Some common steps involved in the extraction of metals from their ores are:

1. Crushing and pulverization.
2. Concentration or dressing of the ore.
3. Calcination and roasting of the ore.
4. Reduction of metal oxides to free metal.
5. Purification and refining of metal.

Crushing and Pulverization

The ore is generally obtained as big rock pieces. These big lumps of the ore are crushed to smaller pieces by using *jaw crushers* and *grinders*. It is easier to work with crushed ore. The big lumps of the ore are brought in between the plates of a crusher forming a jaw. One of the plates of the crusher is stationary while the other moves to and fro and the crushed pieces are collected.

Figure: Jaw Crusher

The crushed pieces of the ore are then pulverized (powdered) in a stamp mill shown in figure. The heavy stamp rises and falls on a hard die to powder the ore. The powdered ore is then taken out through a screen by a stream of water. Pulverization can also be carried out in a ball mill. The crushed ore is taken in a steel cylinder containing iron balls. The cylinder is set into revolving motion. The striking balls pulverize the crushed ore into fine powder.

Figure: Stamp Mill

Concentration or Dressing of the Ore

Generally, the ores are found mixed with earthy impurities like sand, clay, lime stone etc. These unwanted impurities in the ore are called *gangue* or *matrix*. The process of removal of gangue from powdered ore is called *concentration* or *ore dressing*. There are several methods for concentrating the ores. The choice of method depends on the nature of the ore.

Some important methods are:

Gravity Separation (Hydraulic Washing)

In this method, the light (low specific gravity) earthy impurities are removed from the heavier metallic ore particles by washing with water. It is therefore, used for the concentration of heavier oxide ores, like hematite (Fe_2O_3), tinstone (SnO_2), and gold (Au). In this method, as shown in the figure, the powdered ore is agitated with water or washed with a strong current of water. The heavier ore settles down rapidly in the grooves and the lighter sandy and earthy materials (gangue particles) are washed away.

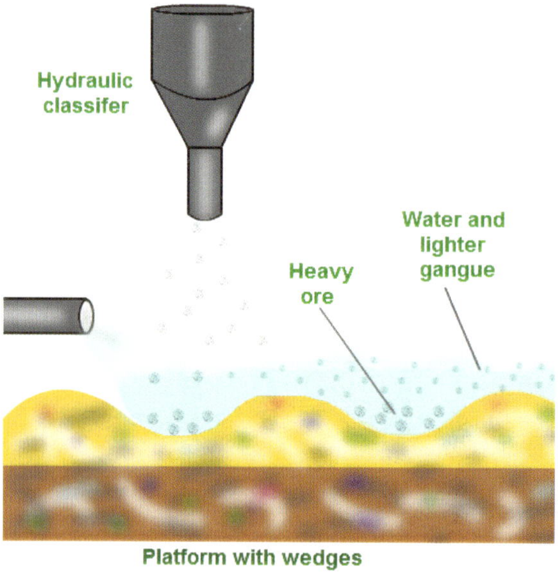

Figure: Gravity separation (Hydraulic washing)

Magnetic Separation Method

By this method, those ores can be concentrated which either contain impurities which are magnetic or are themselves magnetic in nature. For example, the tin ore, tin stone (SnO_2) itself is non-magnetic but contains magnetic impurities such as iron tungstate ($FeWO_4$) and manganese tungstate ($MnWO_4$). The finely powdered ore is passed over a conveyer belt moving over two rollers, one of which is fitted with an electromagnet. The magnetic material is attracted by the magnet and falls in a separate heap. In this way magnetic impurities are separated from non-magnetic material.

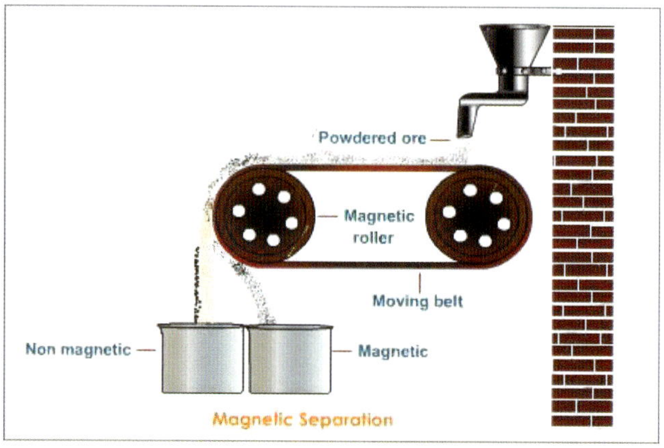

Figure: Magnetic Separation

Froth Floatation Method

This method is especially applied to sulphide ores, such as Galena (PbS), Zinc Blende (ZnS), or Copper Pyrites ($CuFeS_2$). It is based on the different wetting properties of the surface of the ore and gangue particles.

The sulphide ore particles are wetted preferentially by oil and gangue particles by water. In this process, finely powdered ore is mixed with either pine oil or eucalyptus oil. It is then mixed with water. Air is blown through the mixture with a great force. Froth is produced in this process which carries the wetted ore upwards with it. Impurities (gangue particles) are left in water and sink to the bottom from which these are drawn off.

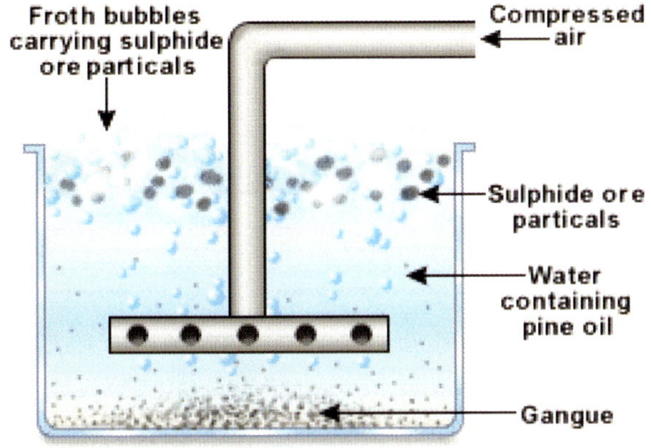

Figure: Froth Floatation

Chemical Method

In this method the ore is treated with a suitable chemical reagent which dissolves the ore leaving behind insoluble impurities. The ore is then recovered from the solution by a suitable chemical method. This is applied for extraction of aluminium from bauxite ($Al_2O_3.2H_2O$). Bauxite is contaminated with iron (III) oxide (Fe_2O_3), titanium (IV) oxide (TiO_2) and silica (SiO_2). These impurities are removed by digesting the powdered ore with aqueous solution of sodium hydroxide at 420 K under pressure. Aluminium oxide dissolves in sodium hydroxide, whereas, iron (III) oxide, silica and titanium (IV) oxide remain insoluble and are removed by filtration.

$$Al_2O_3+6NaOH \rightarrow 2Na_3AlO_3+3H_2O$$

Sodium aluminate (Na_3AlO_3) is diluted with water to obtain precipitate of aluminium hydroxide. It is filtered and ignited to obtain pure alumina.

$$Na_3AlO_3+3H_2O \rightarrow Al(OH)_3+3NaOH$$

$$2Al(OH)_3 \xrightarrow{\Delta} Al_2O_3+3H_2O$$

Calcination and Roasting of the Ore

The concentrated ore is converted into metal oxide by calcination or roasting.

Calcination

Calcination involves heating of the concentrated ore in a limited supply of air so that it loses moisture, water of hydration and gaseous volatile substances. The ore is heated to a temperature so that it does not melt. Two examples of calcination are given below:

1. Removal of water of hydration

   $Al_2O_3.2H_2O \rightarrow Al_2O_3 + 2H_2O$

2. Expulsion of CO_2 from carbonate

   $ZnCO_3 \rightarrow ZnO + 2CO_2$

Roasting

Roasting is a process in which the concentrated ore is heated in a free supply of air at a temperature insufficient to melt it. The following changes take place during roasting:

1. Drying of the ore.
2. Removal of the volatile impurities like arsenic, sulpher, phosphorus and organic matter.

   $4As + 3O_2 \rightarrow 2As_2O_3$

   $S + O_2 \rightarrow SO_2$

   $4P + 5O_2 \rightarrow P_4O_{10}$

3. Conversion of the sulphide ores into oxides

   $2PbS + 3O_2 \rightarrow 2PbO + 2SO_2$

   $2ZnO + 3O_2 \rightarrow 2ZnO + 2SO_2$

Calcination and roasting are generally carried out in a reverberatory furnace or in a multiple hearth furnace.

Reduction of Metal Oxides to Free Metal

This process is carried out after calcination or roasting of ores. In this process called *smelting*, the oxide ores are converted into the metallic state by reduction.

Smelting

Smelting is a process in which the oxide ore in molten state is reduced by carbon or other reducing agents to free metal.

- By using carbon as a reducing agent:

This method is used for the isolation of iron, tin and zinc metals from their respective oxides. The oxide ores are strongly heated with charcoal or coke. Reduction occurs by the action of carbon and/or carbon monoxide which is produced by the partial combustion of coke or charcoal.

$Fe_2O_3+3C\rightarrow 2Fe+3CO$

$Fe_2O_3+CO\rightarrow 2FeO+CO_2$

$FeO+CO\rightarrow Fe+CO_2$

$SnO_2+2C\rightarrow Sn+2CO$

$ZnO+C\rightarrow Zn+CO$

Although the ore has been concentrated in an earlier step, it is still contaminated with some gangue material which is finally removed in the reduction process by the addition of *flux* during smelting.

Flux is a chemical substance which combines with gangue at higher temperatures to form easily fusible material called *slag* which is not soluble in the molten metal. Fluxes are of two types:

- Basic Flux

On heating, lime stone is converted into calcium oxide used as basic flux which combines with acidic impurities like silica in metallurgy of iron and forms fusible calcium silicate ($CaSiO_3$).

$CaCO_3\rightarrow CaO+CO_2$

$CaO+SiO_2\rightarrow CaSiO_3$

(Basic flux) (Acidic gangue) (Slag)

- Acidic Flux

  SiO_2 is used as acidic flux to remove basic impurity of FeO in metallurgy of Cu.

  $SiO_2+FeO\rightarrow FeSiO_3$

  (Acidic flux) (Basic gangue) (Slag)

The fusible slag, such as calcium silicate formed during smelting floats over the molten metal and is thus easily removed. Another advantage is that the slag provides a covering to the molten metal thus preventing it from getting oxidized by air.

- Other reducing agent

Oxide ores which cannot be reduced by carbon or metals which show affinity to carbon by forming metal carbides are reduced by reducing agents like aluminium, sodium, magnesium or hydrogen. *Oxide* like chromium oxide (Cr_2O_3) or manganese oxides (Mn_3O_4) are reduced by aluminium powder is a highly exothermic reaction.

This process is known as Goldschmidt's Alumino-thermite reduction method.

$Cr_2O_3+2Al\rightarrow 2Cr+Al_2O_3+ Heat$

$3Mn_3O_4+8Al\rightarrow 9Mn+4Al_2O_3+ Heat$

Heat is generated in the process due to the formation of Al_2O_3 which is a highly exothermic reaction.

Titanium is obtained by the reduction of $TiCl_4$ (produced by the action of carbon and chlorine on TiO_2) by Mg in an inert atmosphere of argon (Kroll process).

$$\text{Reaction}: TiCl + 2Mg \xrightarrow{\Delta, 1103K} Ti + 2MgCl$$

Titanium can also be obtained by the reduction of TiO2 by sodium

$$TiO_2 + 4Na \rightarrow Ti + 2Na_2O$$

Tungsten and molybdenum can be obtained by the reduction of their oxides by hydrogen,

$$MoO_3 + 3H_2 \rightarrow Mo + 3H_2O$$

- Self-reduction

This is applied to the sulphide ores of copper, mercury and lead. The ores are heated in air, a part of these sulphide ores is changed into the oxide or sulphate which then reacts with the remaining part of the sulphide ore to give the metal and Sulpher dioxide. The reactions showing their extraction are given below:

$$2Cu_2S + Cu_2S \rightarrow 2Cu_2O + 2SO_2$$

Copper glance

$$2Cu_2O + Cu_2S \rightarrow 6Cu + SO_2$$

Copper produced at this stage is called *Blister copper*. The evolution of Sulpher dioxide produces blisters on the surface of solidified copper metal.

$$2HgS + 3O_2 \rightarrow 2HgO + 2SO_2$$

Cinnabar

$$2HgO + HgS \rightarrow 3Hg + SO_2$$

$$2PbS + 3O_2 \rightarrow 2PbO + 2SO_2$$

Galena

$$PbS + 2O_2 \rightarrow PbSO_4$$

$$PbS + 2PbO \rightarrow 3Pb + SO_2$$

$$PbS + PbSO_4 \rightarrow 2Pb + 2SO_2$$

Reduction of Concentrated Ores by other Methods

Some metals cannot be obtained from their ores by using common reducing agents such as C, CO, and H_2 etc. Other methods of reduction are used for such cases.

- Reduction by precipitation

Noble metals like silver and gold are extracted from their concentrated ores by dissolving metal ions in the form of their soluble complexes. The metal ions are then regenerated by adding a suitable reagent. For example, concentrated argentite ore (Ag_2S) is treated with a dilute solution of sodium cyanide (NaCN) to form a soluble complex:

$$Ag_2S+4NaCN \rightarrow 2[Ag(CN)_2]+Na_2S$$

This solution is decanted off and treated with zinc to precipitate silver,

$$2[(CN)_2]+Zn \rightarrow Na_2[Zn(CN)_4]+2Ag(\downarrow)$$

- Electrolytic Reduction

Active metals like sodium, potassium and aluminium etc., are extracted by the electrolysis of their fused (*molten*) salts. For example, sodium is obtained by the electrolysis of fused sodium chloride (Down's process). The s taking place in the electrolytic cell are:

$$NaCl \leftrightarrow Na^+ + Cl^-$$

Na^+ ions move towards the cathode and Cl^- ions move towards the anode. Following s take place at the electrodes:

(At Cathode): $Na^+ + e- \rightarrow Na$ (Reduction)

(Negative electrode) (metal)

(At Anode): $Cl^- \rightarrow Cl + e^-$ (Oxidation)

(Positive electrode)

$$Cl+Cl \rightarrow Cl_2$$

Aluminium is extracted from molten alumina (Al_2O_3) by electrolysis. The melting point of alumina is quite high (2323 K) which is inconvenient for electrolysis. It dissolves in molten cryolite (Na_3AlF_6) at around 1273 k. The s which take place in the cell are:

(At Cathode): $Al^{+3} + 3e^- \rightarrow Al$ (Reduction)

(Negative electrode) (metal)

(At Anode): $C + 2O^{2-} \rightarrow CO_2 + 4e^-$ (Oxidation)

(Positive electrode)

Purification and Refining of Metal

Except in the electrolytic reduction method, metals produced by any other method are generally impure. The impurities may be in the form of (i) other metals (ii) unreduced oxide of the metal (iii) non-metals like carbon, silicon, phosphorus, Sulpher etc. and (iv) flux or slag. Crude metal may be refined by using one or more of the following methods:

Liquation

Easily fusible metals like tin, lead etc. are refined by this process. In this method, the impure metal is poured on the sloping hearth of a reverberatory furnace and heated slowly to a temperature little above the melting point of the metal. The pure metal drains out leaving behind infusible impurities.

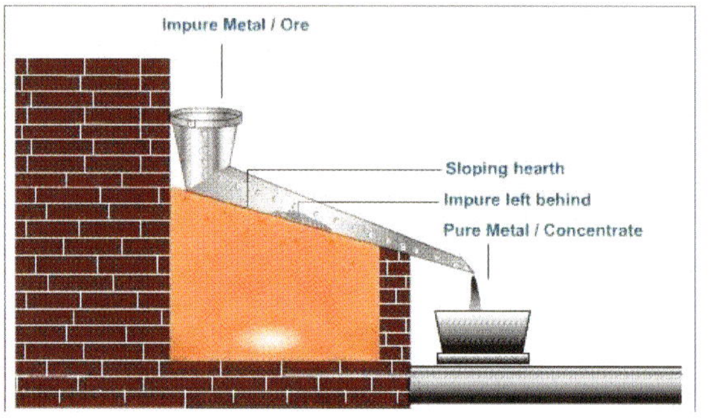

Figure: Liquation

Poling

Poling involves stirring the impure molten metal with green logs or bamboo. The hydrocarbons contained in the pole reduce any metal oxide present as impurity. Copper and tin are refined by this method.

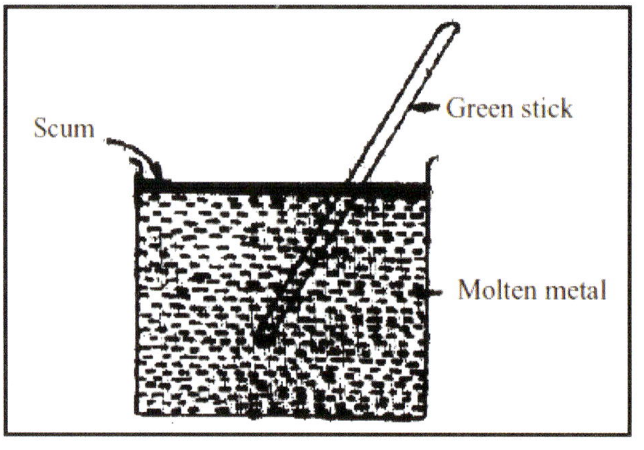

Figure: Poling

Distillation

Volatile metals like zinc and mercury are purified by distillation. The pure metal distils over, leaving behind non-volatile impurities.

Electrolytic Refining

A large number of metals like copper, silver, zinc, tin etc. are refined by electrolysis. A block of impure metal is made the anode and a thin sheet of pure metal forms the cathode of the electrolytic cell containing suitable metal salt solution which acts as an electrolyte. On passing current, pure metal deposits at the cathode sheet while more electropositive impurities are left in solution. Less electropositive metals do not dissolve and fall away from the anode to settle below it as *anode mud.*

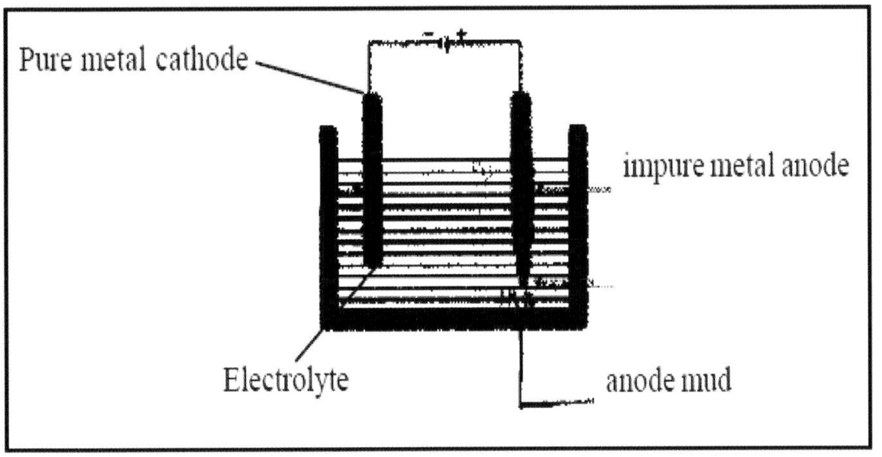

Figure: Electrolytic Refining

For example, in the electrolytic refining of crude copper (blister copper), a large piece of impure copper is made anode and a thin piece of pure copper is made the cathode. An acidified solution of copper sulphate is used as an electrolyte. On passing an electric current of low voltage through the solution copper (II) ions obtained from copper sulphate solution go to the cathode where they are reduced to the free copper metal and get deposited.

(At Cathode): $Cu^{+2}+2e^{-}\rightarrow Cu$ (Reduction)

(Negative electrode) (metal)

An equivalent amount of the metal from the anode dissolves into the electrolyte as Cu^{+2} ions.

(At Anode): $Cu\rightarrow Cu^{+2}+2e^{-}$ (Oxidation)

(Positive electrode)

As the process goes on, anode becomes thinner while the cathode becomes thicker. The impurities like silver, gold settle down at the bottom of the cell as *'anode mud'*.

Examples

Extraction of Cu in Pyrometallurgical Process

- Details about of Cu:

  Atomic Number: 29

  Atomic Weight: 63.54 amu

  Density: 8.94 gm/cm ^{3}

  Melting Point: 1083° C

  Boiling Point: 2595 ^{o} C

- Common Minerals of Cu:

Copper Oxides		Copper Sulphides	
Name	Chemical Formula	Name	Chemical Formula
Cuprite	Cu_2O	Chalcosite	Cu_2S
Chrysocolla	$CuSiO_3.2H_2O$	Chalcopyrite	$CuFeS_2$
Malachite	$CuCO_3.Cu(OH)_2$	Bornite	Cu_5FeS_4
Azurite	$2CuCO_3.Cu(OH)_2$	Covellite	CuS

- Extraction of Cu from Sulphide Ores:

The steps involved in extraction of Cu by conventional routes as concentration, roasting, smelting, converting and refining at the left side of the flow sheet drawn in figure and newer route at the right side of the flow sheet.

Conventional Route

Concentration: A naturally occurring Cu sulphide contains 0.5-2 % of Cu. To recover this, first we go for crushing and grinding to liberate sulphide grains from the gangue of average 40 μm sizes.

Then go for froth floatation, where addition of CaO as P^H controls and Xanthate as collecter, where we separate Cu sulphide from a mixture of Cu Sulphide, Lead Sulphide and Zinc Sulphide. The Cu Sulphide contains 15-35% Cu, 15-35% Fe, 25-35% S, 3-15% gangue.

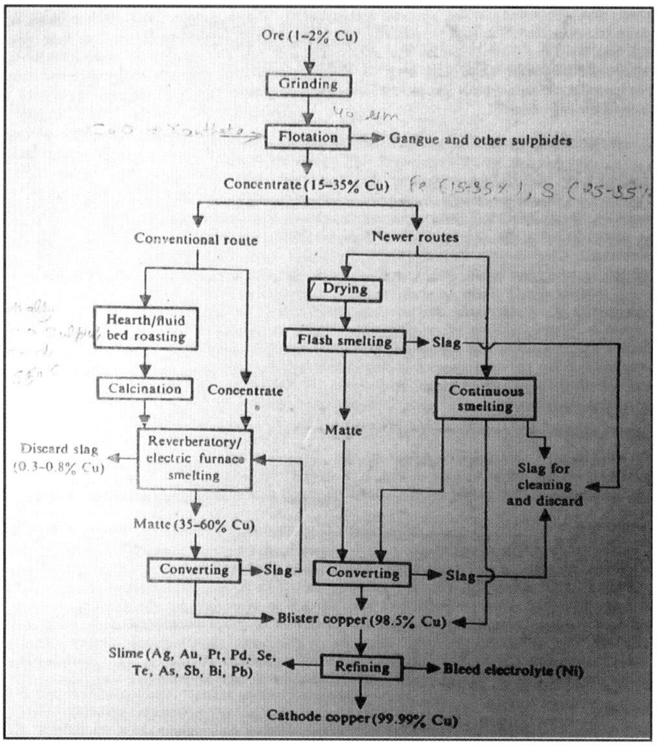

Figure: Pyrometallurgical Extraction of Cu from Sulphide Ores

Roasting: The purpose of roasting to partially oxidized the iron sulphide present in Cu Sulphide which from slag in next i.e. smelting. Here the concentrate contains high iron sulphide and low Cu Sulphide content (<25%).

In conventional plant, roasting carried out in multiple hearth roaster with gradually downward movement of feed having gradual increase in temperature of each hearth by fuel firing as 550 ° C. Where as in newer route, roasting carried out in fluidized bed roaster at about 550 ° C. The product is SO_2 gas which finally used for manufacturing for H_2SO_4.

The principal reactions takes place in roasted calcine contains Cu Sulphide and Fe oxides of Fe and mixed sulphates of Fe and Cu given as

$$2CuFeS_2+6.5O_2 \rightarrow 2CuO+Fe_2O_3+4SO_2$$

$$CuFeS_2+4O_2 \rightarrow CuSO_4+FeSO_4$$

$$2CuFeS_2+O_2 \rightarrow Cu_2S+2FeS+3SO_2$$

$$2CuFeS_2+4.5O_2 \rightarrow 2Cu_2S+Fe_2O_3+3SO_2$$

$$2CuFeS_2+7.5O_2 \rightarrow CuO.CuSO_4+2FeSO_4+SO_2$$

$$3CuFeS_2+9.5O_2 \rightarrow 3CuO+Fe_3O_4+6SO_2$$

$$6CuFeS_2+13O_2 \rightarrow 2Cu_2S+2Fe_3O_4+9SO_2$$

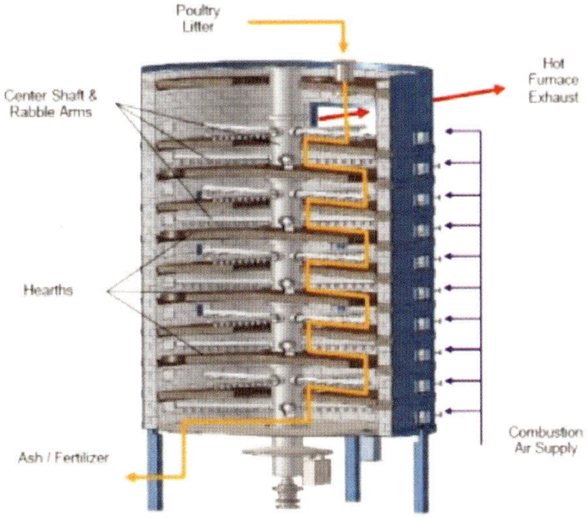

Figure: Multiple Hearth Roaster

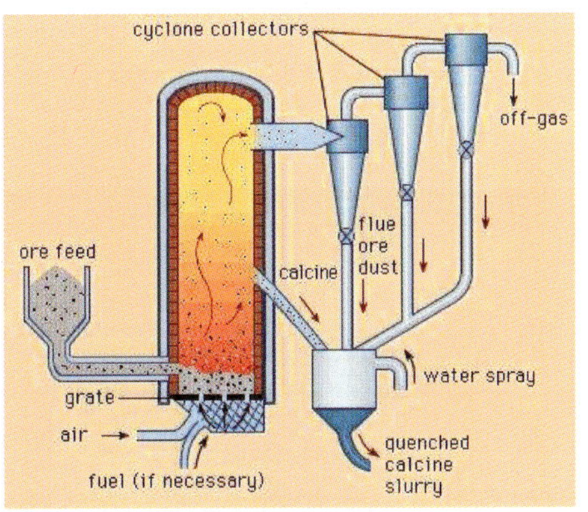

Figure: Fluidized Bed Roaster

Smelting

It seeks to separate the metal sulphides in an ore concentrate or calcine from the gangue by charging the furnace with suitable charge along with flux at about 1250 ° C. During smelting two separate layers are formed, as slag and matte layer having specific gravity 2.8-3.8 gm/cm^3 and 5.0-5.5 gm/cm^3 respectively. If gangue and iron oxides present in the concentrate/calcine to be smelted combine during smelting to form iron silicate slag. If iron oxide insufficient the lime add as flux to form an iron calcium silicate slag. The melting point of slag approximately 1150 ° C and for fluid slag with a low Cu content, the temperature at about 1250 ° C.

Exchange reaction takes place between oxides and sulphides of Cu and FeS present in furnace charge.

$$2CuO+4FeS \rightarrow 2Cu_2S+4FeO+SO_2$$

$$2CuSO_4+2FeS \rightarrow Cu2S+2FeO+3SO_2$$

$$Cu2O+FeS \rightarrow Cu2S+FeO$$

These reactions occur because O_2 has a greater affinity for Fe than Cu in Ellingham diagram. The unoxidised iron sulphide reduces the higher oxides of iron to ferrous oxide as

$$10Fe_2O_3+FeS \rightarrow 7Fe_3O_4+SO_2$$

$$3Fe_3O_4+FeS \rightarrow 10z+SO_2$$

Smelting determine the extent of oxidation of FeS in charge to slag object is to produce a matte contains 35-45% Cu, 20-22% S, 25-35% Fe. Smelting not only minimize the loss of Cu to slag, but also provides a matte with a sufficient quantity of iron sulphide for use in the next stage as converting.

Smelting conventional route occur in reverberatory furnace and electric furnace, where in newer route flash smelting and continuous smelting takes place. The conventional route, electric furnace because a large volume of combustion gas generation avoided and facilities both recovery of so_2 and cleaning of furnace gases.

Converting: The purpose of converting is to remove Fe, S, and other impurities from matte. Generally side blown converter used.

The capacity of converter is 100-200 tons, 4m diameter and 9 m in length. Lined with a layer of Chrome-magnesite refractory about 40 cm thick. Oxidizing atmosphere inside the converter by supply of O_2 enriched air through the tuyere of 5 c.m. with 40 in number. The total volume of gas is 600 m^3/min. The product is blister Cu and slag. So, the Cu converting namely slaging stage and blister formation stage.

(i) Slaging Stage: The iron sulphide present in matte is oxidized and oxide is slaged out by addition of siliceous flux. The reactions are:

$$2FeS+3O_2 \rightarrow 2FeO+2SO_2$$

$$2FeO+SiO_2 \rightarrow 2FeO.SiO_2$$

Mainly three process are done as

- Adding freshly obtain matte.
- Blowing of air.
- Slag gets skimmed off by tilting.

The slag contains 2-9% Cu, 40-50% Fe, 20-30% SiO_2, and 1-5% (CaO+MgO). Conventionally, Cu is recovered from the slag produce in converter by transferring the slag in molten state to a smelting furnace, where in case of newer process, metallic values of the slag recovered by employing either an electric furnace with provision for cleaning the slag or by slow cooling of slag followed by grinding and floatation.

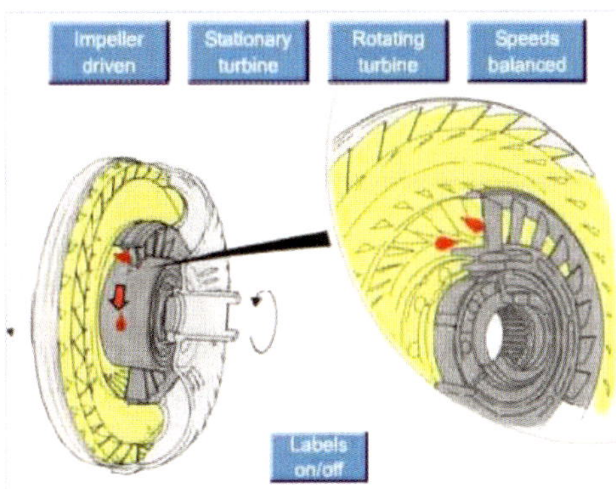

Figure: Converter Operation

(ii) Blister Formation Stage: After slagging converter contains Cu_2S known as white metal because of its appearance. In blister Cu formation stage, Cu_2S is oxidized to form Cu by combination of reaction.

$$2Cu_2S+3O_2 \rightarrow 2Cu_2O+2SO_2$$

$$Cu_2S+2Cu_2O \rightarrow 6Cu+SO_2$$

(Overall): $3Cu_2S+3O_2 \rightarrow 6Cu+3SO_2$

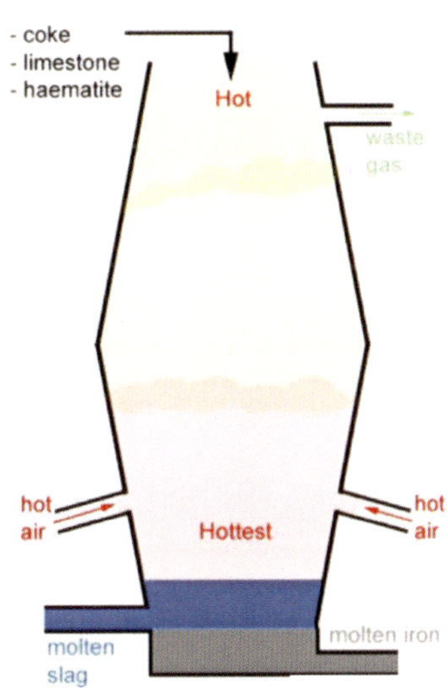

Figure: Operating Principles of Side blown Converter

When white metal (Cu_2S) is oxidized the S content in bath decrease about 19.5% and hence, the bath split into two layers as:

- Top sulphide layer of less dense.
- Bottom Cu layer of more dense contain about 1.2% S.

As oxidation continues the sulphide layer (liquid b) decrease and Cu layer (liquid c) increase. The relative volume of two phases determined by Lever rule.

The blister Cu contents 0.02-0.05% S along with 0.2-0.5% dissolved O_2.

Refining

The purpose of refining to get Cu extraction is twofold as

- First, to obtain metal in pure form.
- Second, to recover precious metals containing in blister Cu produced.

(i) Fire Refining: The purpose is to remove S from liquid blister Cu as SO_2 by oxidation with air and eliminates O_2 by introducing hydrocarbons. In this case, the refining is done in reverberatory furnace of 400 ton of capacity contains blister Cu get oxidized to recover Cu removing the impurities such as S, Fe, Se, Zn by converting its corresponding oxides and then skimmed off. But, some Cu also in the form of oxides.

To prevent Cu loss poling with green branches used to reduce Cu_2O using hydrocarbon or some other reducing gases. In this case, the purity of Cu obtained 99.97%.

Fire refining is also done in rotary type refining furnace, where blister Cu directly treated by blowing air.

(ii) Electrolytic Refining: The purpose is to further refined the fire-refined Cu by electrolysis. The electrolysis done in a electrolytic refining tank made of concrete or wood of 3-5 m deep and utilization minimum space with maximum cathode and anode area. The electrolyte is $CuSO_4$, H_2SO_4, some glue and alcohol at temperature 50-60° C.

Cu transferred from crude anode to pure cathode. Impurities in blister Cu such as Fe, Co, Ni, Se, Te etc. go into the solution and precious metals get collected below anode known as anode slime.

(ii) Newer Process.

Flash Smelting

Process: Enriched preheated air or pure O_2 used to increase combustion rate and autogeneous smelting. The gases coming out rich of SO_2 due to high combustion rate and used for H_2SO_4 manufacturing.

Process is autogeneous provide exothermic heat. Air used as oxidant to preheated. The composition of concentrate used in flash smelting has Chalcopyrite ($CuFeS_2$) 66%, Pyrite (FeS_2) 24%, gangue (SiO_2) 10%. Whereas the Cu matte contains 70% Cu, 8% Fe, 22% S, slag contains Fe 40% at furnace temperature 1300° C.

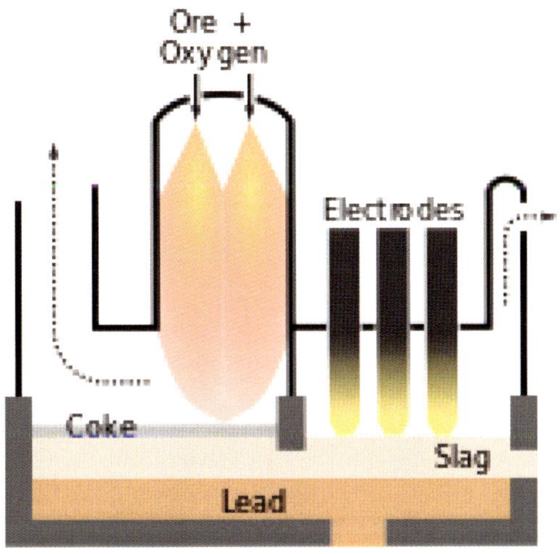

Figure: Flash Smelting

Main reactions of flash smelting of Cu concentrate are mentioned below:

$$CuFeS_2 + 0.5O_2 \rightarrow 0.5Cu_2S + FeS + 0.5SO_2$$

$$CuFeS_4 + 0.5O_2 \rightarrow 2.5Cu_2S + FeS + 0.5O_2$$

$$CuS + 0.5O_2 \rightarrow 0.5Cu_2S + 0.5SO_2$$

$$FeS_2 + O_2 \rightarrow FeS + SO_2$$

$$FeS + 1.5O_2 \rightarrow FeO + SO_2$$

$$FeO + 0.16O_2 \rightarrow 0.33Fe_3O_4$$

$$FeO + 0.5SiO_2 \rightarrow 0.5Fe_2SiO_4$$

$$Cu_2S + 1.5O_2 \rightarrow Cu_2O + SO_2$$

Continuous Smelting

It encompass smelting and converting in a single vessel i.e. Cu concentrate charged at one end and Cu metal withdrawn continuously at the other end.

Mainly three are three processes, given below:

(a) WORCRA. (b) Noranda. (c) Mitsubishi.

WORCRA Process

This name divided as the first 3 alphabets stands for the developers and last 3 alphabets for the place.

Features

- Counter current movement of gas and concentrate. So, continuous production of blister Cu.
- Directly blister Cu i.e. metal instead of matte form.
- Combine smelting and converting.
- The heat required for reaction directly obtained as the reaction is exothermic.
- Counter movement cause continuous production of H_2SO_4 due to continuous extraction of gas.
- Cu% continuously obtains from slag by means of cleaning operation.

Process

The process combines 3 different operations in a single furnace as:

- Continuously smelting
- Continuously converting
- Continuously slag cleaning by conditioning and settling

Efficiency: It increases by means of counter current movement increase the reaction surface area in the smelting and converting zone. Hence, effective removal of impurity occur i.e. mainly Fe due to counter movement of slag and matte. As a result, Cu gets reverted back to matte and obtain.

Advantage

- Continuous process
- Capital cost low
- Concentrate passes large surface area. Hence accelerate the reaction

Disadvantage

- Not durable
- Operating cost is high

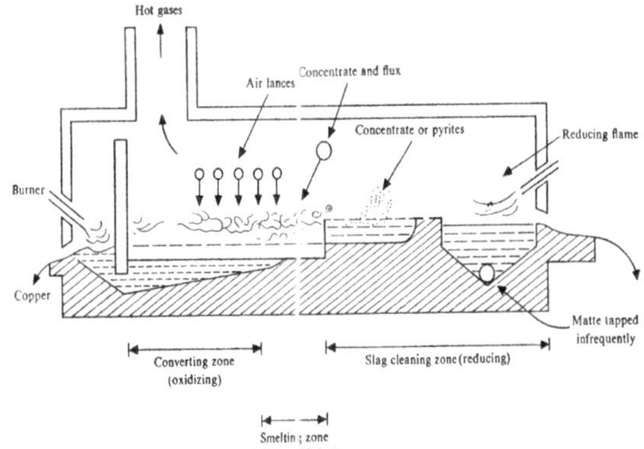

Figure: Vertical Sectional Diagram of Straight form of WORCRA reactor

Noranda Process

Principle: In this process, high grade Cu matte directly forms from the sulphide by means of air blown through the tuyeres to oxidize. The Cu or matte collected at the tap hole where the slag collected at the other tap hole at the slag end.

In this process, the slag contains high% Cu compare to WORCRA process. Mainly 3 layers present in the product as:

- Cu – First layer.
- Matte – Second layer.
- Slag – Third layer.

Condition

- If given air is more than the stoichiometric amount of air required for oxidation, then matte level decreases and Cu level increases.
- If insufficient air required for stoichiometric amount then unoxidised iron and S tends to combine with Cu to form matte. So matte level increases and Cu level decreases.
- If air supply is equal to the stoichiometric amount then both matte and Cu level get increases.

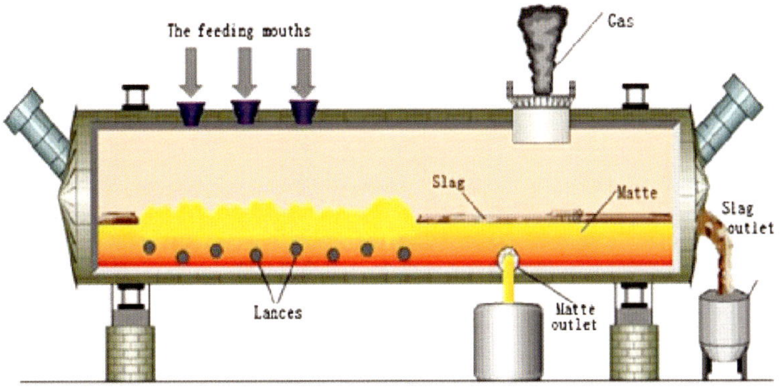

Figure: Schematic diagram of Noranda Process

Mitsubishi Process

Principle: There are three furnaces such as smelting, slag cleaning and converting furnace are connected in a cascade manner. The product of one furnace goes to next furnace for next operation by means of gravity force.

Process:

- First, in smelting furnace (wet concentrate + flux + air + O_2) is smelted to produced matte of 60-65% Cu and rest is slag.
- Secondly, both matte and slag goes into slag cleaning furnace where slag get discarded and matte goes to next furnace operation.

- Thirdly, in converting furnace matte oxidized to blister Cu by blow of O_2 enriched air and limestone add as slag. So slag discarded as lime ferrite. Blister Cu produced of low % S and hence, obstruct the transfer of Cu to lime ferrite slag.

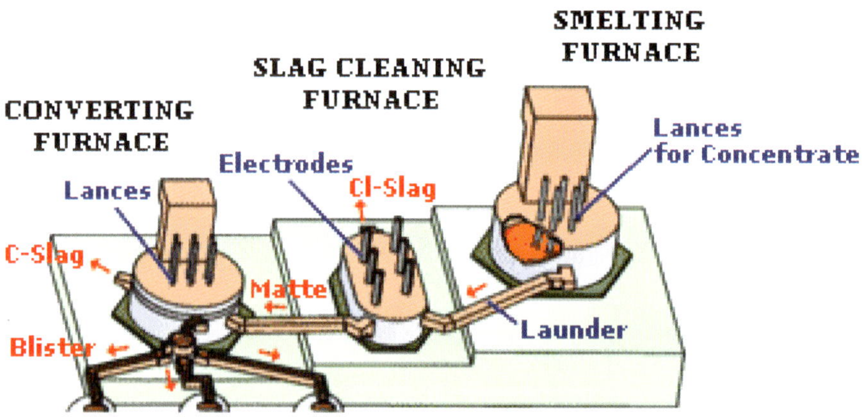

Figure: Mitsubishi Continuous Smelting Process

Area of Extraction

- HCL (Hindustan Copper Ltd.)
- Khetri mines
- Kholihana mines

Uses

- Conducting element.
- Alloying addition.

Extraction of Zn by Pyrometallurgical Process

Details of Zn

Atomic Number: 30

Atomic Weight: 65.38 amu

Density: 7.13 gm/cm ^{3}

Melting Point: 419.5°C

Boiling Point: 910 $^\circ$C

Common Minerals of Zn

Zinc Oxides		Zinc Sulphides	
Name	Chemical Formula	Name	Chemical Formula
Zincite	ZnO	Sphalerite	ZnS

Franklinite	[ZnO(Fe, Mn)$_2$O$_3$]		
Willemite	Zn_2SiO_4		
Smith Sonite	$ZnCO_3$		

Extraction of Zn from Sulphide Ores

Mainly five important processes are there in extraction of Zn. Given as follows:

1. Horizontal Retort
2. Vertical Retort
3. Electrothermic
4. Elcctrolytic
5. Imperial Smelting

The first 3 processes are retort process. Horizontal retort made a major contribution to Zn production. Electrolytic process gives high degree purity at about 99.95%. In this case Si rectifiers incorporated in electrolytic cells to reduce power consumption and higher current density achieve.

Recently imperial smelting process (ISP) used, in which ZnO get reduces by use of molten lead to give Zn vapour which on condensation gives Zn.

$$ZnS+1.5O_2 \rightarrow ZnO+SO_2$$

$$ZnO+CO \rightarrow Zn+CO_2$$

$$C+CO_2 \rightarrow 2CO$$

Flow Sheet Explanation

The technique of smelting and converting not applicable for Zn extraction because ZnS not melt even at 1500° C. Zn% 55 in concentrate treated under roasting, then it go for sintering where roasted oxide ground and agglomerates then go under C reduction to give Zn in vapour form. For volatisation hearth roaster used instead of other.

Initial stage roasting i.e. vitalization a hearth roaster employs. Then remaining PbS removed during sintering rather than high temperature. In order to prevent distillation of Pb present in Zn concentrate to obtain high grade Zn with low Pb in retort distillation.

In most cases, fluid bed roaster allotted instead of hearth roaster because of following advantages

- Autogeneous.
- Good control over high through put of Zn calcine.
- Generate SO_2 suitable for acid production.

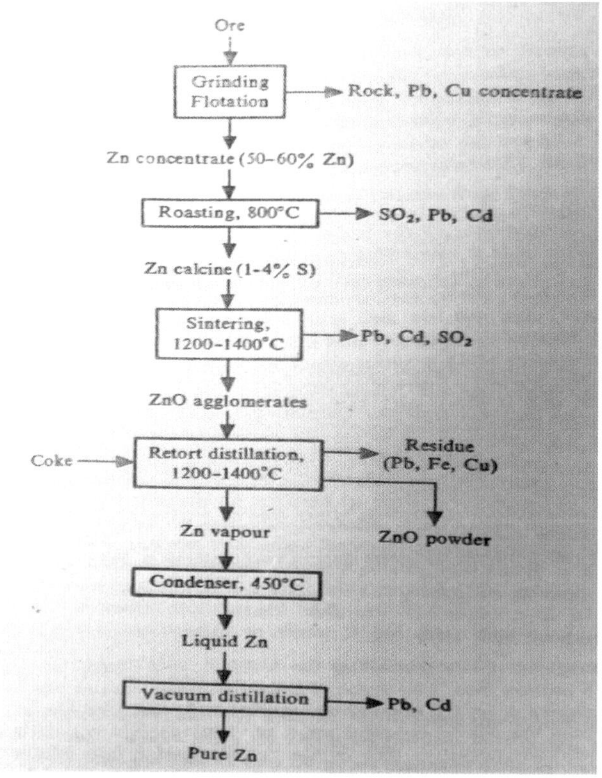

Figure: Extraction of Zinc by Reduction of ZnO by Carbon

Horizontal Retort Reduction

It is a 2m long, 0.25 m dia., 5-6 cm thick gas fired retort furnace made of clay used instead of metal retort because most metals gives alloy when heated with Zn. Here, first loose mixture of ZnO presence of coke breeze charged at 1400° C to volatize Zn which on condensation gives Zn.

Vertical Retort Reduction

It is a 10 m height, 0.7 m dia., and 30 cm thick of capacity 7-10 tons Zn per clay made of SiC because has high conductivity approximately 5 times compare to horizontal retort.

Electrothermic Reduction

It occurs in a vertical retort 15 m high, internal dia. of 24 m about 100 tons Zn per day. Electrodes of graphite introduce through SiC walls at two places near the bottom and at a zone of 9 m from the bottom. The gas liberate due to the reduction of charge contains 40-45% Zn, 45% CO, 5-8% H_2, 5-10% N_2. Avoid formation of blue oxide of Zinc (ZnO+Zn) quick condensation takes place. Residue contains 3-4% Zn. Zn recovery over 95%.

Zn product obtained in foregoing retort is known as Spelter.

Three distinct layers as

- Bottom layer is molten lead contains some Zn of 1.5%.

- Top layer is Zn contains some lead of 0.8%.
- Intermediate layer of Zn and Fe (15-20)% called hard metal.

Zn layer further refined by fractional distillation because wide difference in boiling point of Zn (907° C), Cd (780° C), and Pb (1620° C) facilities the separation of one metal from the other.

Electrolytic Reduction

In this case, when Zn immersed in their salt solutions, rapidly establish a reversible potential. Since the activation over potential is low, the displacement behavior exhibits a good correlation with the electrochemical series of standard potentials.

Imperial Smelting Process (ISP)

The process inside the furnace based on reaction i. e. reduction of ZnO by C in the imperial blast furnace gives rise to Zn in vapour form which gets condensed by using molten lead.

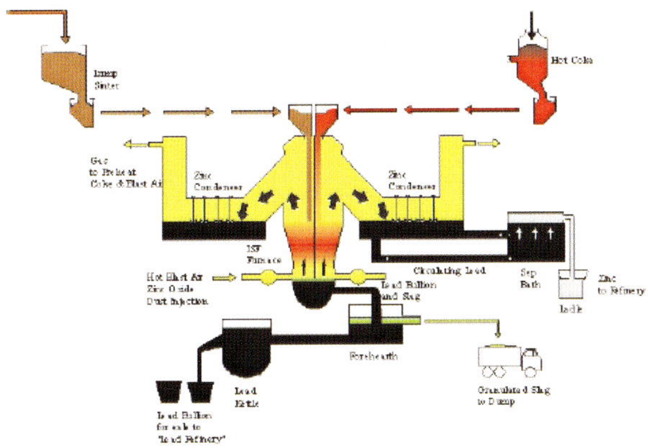

Schematic representation of an ISP furnace

Imperial Smelting Blast Furnace

Principle: The process is carried out by using counter current principle, where the blast and preheated air given to the furnace through the tuyere and the preheated coke get charged from the top.

- ISP furnace is of square cross section consists of water jacket brick lined shaft.
- Feed is given from the top of the furnace at positive pressure where we introduced preheat-ed coke and sinter through a double bell charging system.
- ISP consists of two condensers.

Function of Condenser: These are used mainly for two purposes as

- Cooling of furnace gases resulting due to condensation of Zn.
- Dissolving condensed Zn in molten Pb.

The smelting reactions takes place in ISP are

$$\text{Reaction}: C + \frac{1}{2}O_2 \rightarrow CO$$

$$C+O2 \rightarrow CO2$$

$$C+CO2 \rightarrow 2CO$$

$$ZnO+CO \rightarrow Zn+CO2$$

$$PbO+CO \rightarrow Pb+CO2$$

Carbon does not reduce ZnO until 1120° C, since at this temperature Zn gets vaporized. After reduction complete the molten slag and Pb are taken out from the furnace hearth at the melting point of the slag. Hence, Zn gets collected. So the reaction may backward as due to the difference.

Lead Bullion (Rich 70-80% Pb): It is the mixture of Pb and precious metal, where preheated coke 800° C and sinter 21% Pb are charged from the top of ISP furnace. So at bottom collect precious metals such as Au, Cu etc. and this liquid is known as Lead Bullion.

Advantages of ISP

- Economical production of Zn.
- Efficiency overall high but Zn recovery expensive.
- Entire amount of Au, Sb etc. recover.
- Capacity large.
- Complete mixed charge of Zn and Pb simultaneously charged and recovered.
- Operational cost is low.
- No additional C required.

Figure: Zinc Recovery System

Zn from Lead Slag by Slag Fuming Process

The process initiated by Consolidated Mining Smelting Corporation (COMINCO) of Trial, Canada.

Process: The slag produce in lead bath or furnace are 15-18% Zn. Molten slag from lead blast furnace together with cold slag from dump get transformed by ladle and charged to the furnace. Then the treatment growing of molten slag by treated with mixture of pulverized fuel and air through tuyere. Hence, Zn get volatized and slag get solidified by means of water jacket in all sides of furnace except roof as roof is cover with fire bricks.

So, due to mixture of pulverized fuel and air through tuyere the reactions occur as

In bath,

$$2C + \frac{3}{2}O_2 \rightarrow CO + CO_2$$

$$CO + ZnO \rightarrow Zn + CO_2$$

Above bath,

$$Zn + \frac{1}{2}O_2 \rightarrow ZnO$$

$$CO + \frac{1}{2}O_2 \rightarrow CO_2$$

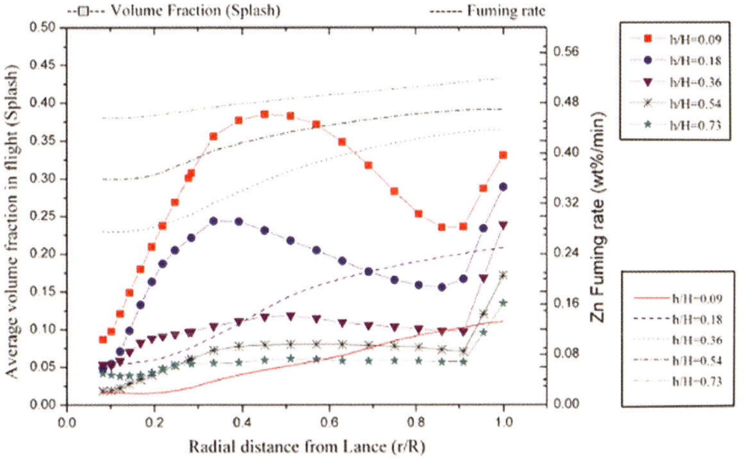

Figure: Zinc Elimination during Slag Fuming

The excess heat gets collected by boiler located in gas collection system and the control of coal to air ratio determine the extent of heat released and maintained the bath temperature at about 1200° C.

First, the elimination of Zn is slow but reaching 1000° C get accelerate so Zn% 2.5%. Whereas Zn 1% in slag product contain 93-94% Zn. Fume leaves furnace at 1100° C by combustion which effected by water cooling tower at 200° C.

Then fume enters to big house where Zn collected and dispatched to Zn smelter on an electrolytic plant.

Uses or Applications

- Zn extensively used as a protective coating for steel (Galvanization).
  - Restrict atmosphere corrosion by impervious basic $ZnCO_3$ layer.
  - Zn is more electropositive.
- Fabrication of Cu-Zn alloys i. e. brasses.
- Spraying: Zn used as for spraying in comparison to other metal on that metal which has low melting point.
- Rolled Zn: Usual method of cold working. Zn is rolled to sheet, plate, and strip. Where, sheet plates are rolled from 98.5% Zn.
- Pigment: Zn in the form of its oxides used in manufactured of paints.
- Alloys: Mainly Cu-Zn alloy produce of 30-37% Zn which is much less plastic when cold and worked about 500°C which mainly used as die casting alloy.

Area of Extraction

- Zawar Mines (Rajstan).
- HZL (Hindustan Zinc Ltd.).
- COMINCO-BINANI at Kerela.

Extraction of Pb by Pyrometallurgical Process

Details of Pb

Atomic Number: 82

Atomic Weight: 207.21 amu

Density: 11.34 gm/cm ^{3}

Melting Point: 327.40°C

Boiling Point: 1737°C

Common Minerals of Pb

Pb Oxides		Pb Sulphides	
Name	Chemical Formula	Name	Chemical Formula
Cerrusite	$PbCO_3$	Galena	PbS
Anglesite	$PbSO_4$		

Extraction of Pb

Production of Pb Bullion

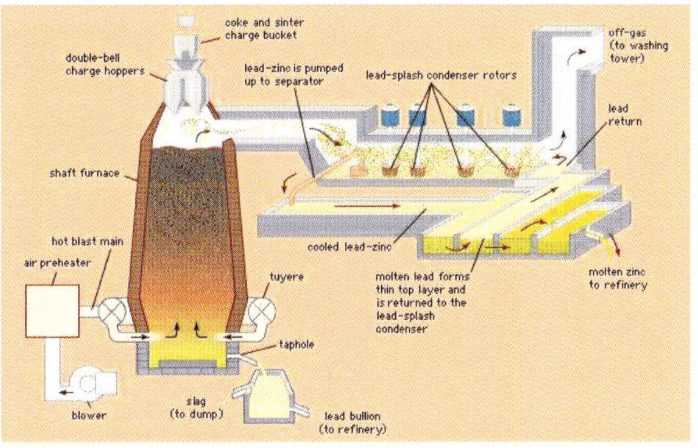

Figure: Flow Sheet of Production of Lead Bullion

Complete Description of Flow Sheet (Extraction of Lead Bullion)

The concentrate roasted to oxides which get reduced by C at 800° C above to fuse PbS, which occur in Dwight-Lloyd sintering machine in which not only S eliminate but also produce sinter with liquidize properties. In this S% reduce from 16-18% to 1-2%.

Now, up draught is more preferred compare to down draught because of

- Higher capacity.
- Elimination of precipitate wind box Pb.
- Production of high grade SO_2 gas.

In, up draught air is admitted from perforated bottom gate. Heat liberated due to roasting of Galena ore out of temperature below fusion temperature.

$$2PbS + \frac{7}{2}O_2 \rightarrow PbO + PbSO_4 + SO_2$$

Hence, free Pb gravitated to the wall of furnace, and was trapped out at working temperature 1000° C. Thick crust allow to accumulate slowly down even stopped reaction by the action of the air blast.

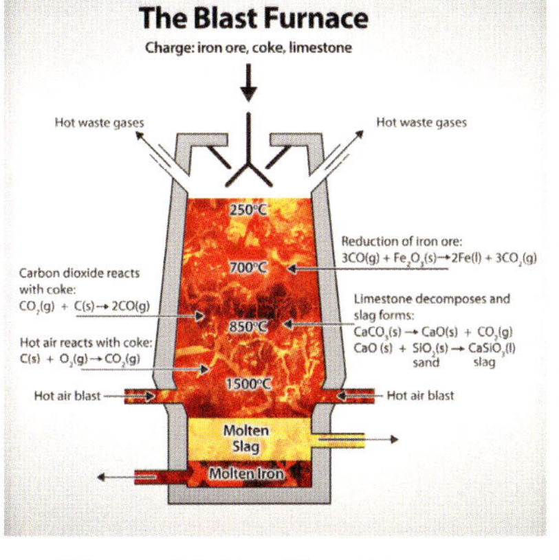

Figure: Molten Blast Furnace

Older Blast Furnace Reduction: Furnace decrease in area from throat downward a minimum being attend end at tuyere of area 4-5 m^2. Blast pressure increases resulting in volatization of lead. As increase volume of blast lead volatization decreases.

Siphon top: A device connects molten Pb in the crucible with a bowl shape which keeps the level of Pb in the crucible constant by withdrawn from the top.

Bosh: Purpose is to retard the descent of charge in front of tuyere as material melts away.

Newer Blast Furnace Reduction: Furnace area increases from the throat to downward at the end of tuyere in this case following advantages are occur:

- Gases more evenly through charge.
- Decreasing their velocity.
- Decrease the amount of dust.
- Here scrap used for reduction of unreduced PbO.

PbO also helps in reducing PbS left unroasted in the sinter machine and hence, combine with Silica (SiO_2) in charge to form $2FeO.SiO_2$. Hence, lower the melting point of slag and increase the fluidity of metal layer.

$$PbSiO_3 + Fe \rightarrow FeSiO_3 + Pb$$

Finally helps to reducing the Pb as

$$PbO + Fe \rightarrow FeO + Pb$$

Disadvantage of Newer Process:

- A thick mushy layer form in between matte and slag which interfering into the separation.
- A wide variation in temperature as 200° C at top to 1200° C at tuyere.

In blast furnace reduction produce 4 distinct layers on the basis of their specific gravity as

- First layer – Slag Layer – Specific gravity 3.6.
- Second layer – Matte Layer – Specific gravity 5.2.
- Third layer – Speiss Layer – Specific gravity 6.
- Forth layer – Lead Bullion/Base Bullion.

Refining of Pb Bullion to Extract Pb

Liquation: Refining technique for removing impurities from metal based on the factor such as difference in melting point of alloy components. So as a number of layers formed on the basis of their difference in density.

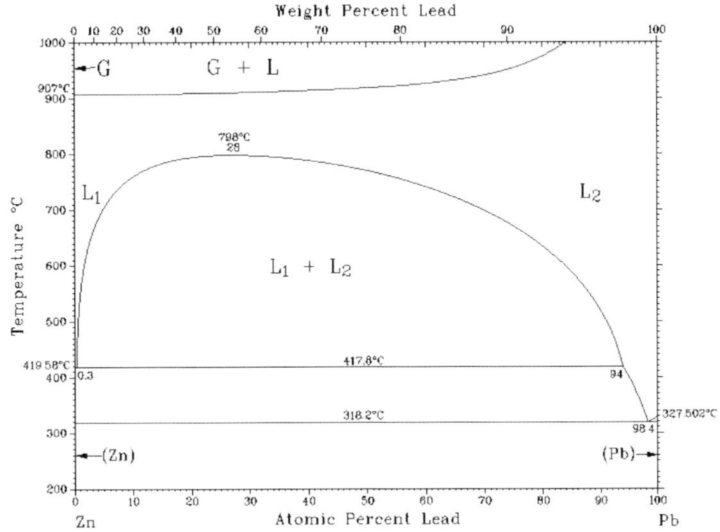

Figure: Lead Zinc Phase Diagram

The crude Zn containing of 5% Pb refined in a reverberatory furnace by slowly cooling of molten matte.

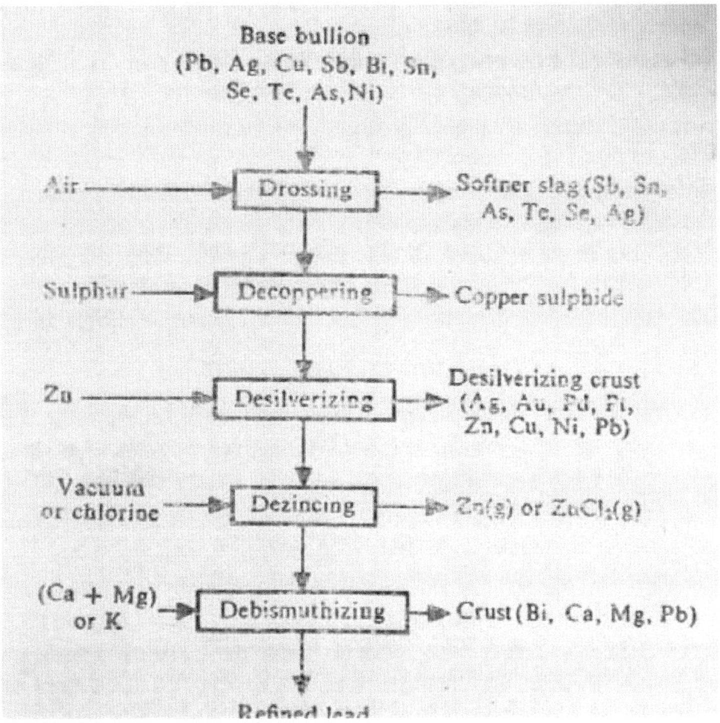

Refining of Lead Bullion

Drossing: It is the combine process of liquation and oxidation in reverbaratory furnace at around temperature 350° C at which many metals oxidized and skimmed off subsequently increase to (500-550)°C. At that temperature Cu get eliminates using S in the form of Cu_2S and at last softener slag (Sb, Sn, As, Te, Se) are removed.

Desilverising of Pb (PARKE'S Process)

The process of desilverising of Pb by Zn addition followed by cooling for Ag recovery is known as desilverising of Pb or PARKE's process.

In this process, Zn is added in molten Pb because of high affinity of Zn for Ag and Au than Pb. Then two liquid layers such as:

- Liquid Zn at the upper layer having Zn% of 0.6.
- Liquid Pb at the bottom layer having Pb% of 1.5.

As Zn layer (first layer) has higher melting point. So, solidify first.

Dezincing

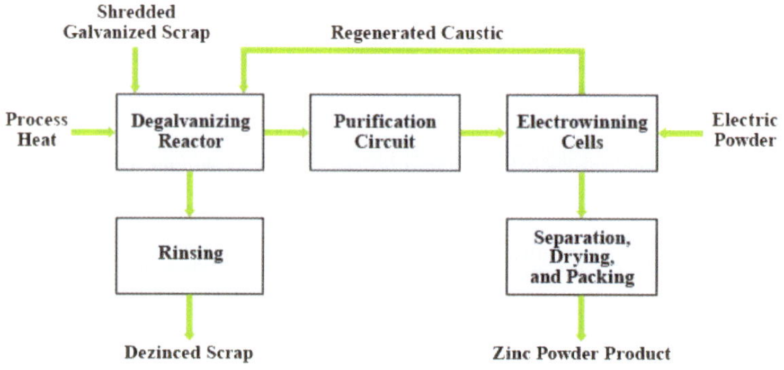

Figure: Dezincing Process

As Cl_2 reacting with Zn to give $ZnCl_2$ in an enclosed chamber cause left of Pb and falling on bath looks grayish scum surface (layer of bubble formation) which gets skimmed off.

At the end point, all Zn chlorinated and white crystal of $PbCl_2$ appeared on surface. Vacuum dezincing desilverised lead pump to kette i. e. bath enclosed by steel ball shaped with a stirrer and vacuum pump attached. As temperature rose to 540° C vacuum distillation carried out in reduction of Zn content from 0.6 to 0.002% in lead.

Debismuthizing

The base bullion consist high Bi% which get separated by addition of Ca and Mg to forms an insoluble intermetallic compound of Bi.

Refining of Lead Bullion to Get Pb using Electrolytic Reefing

In the electrolytic cell, the pure Pb obtained by taking base or Pb bullion at anode where, electrolyte consists of a solution of $PbSiF_6$ (Lead Flouro Silicate) in H_2SiF_6 (Hydrogen Silicic acid). In this case, H_2SiF_6 get acted on white Pb. So, granulated Pb obtained to give $PbSiF_6$ in the next step.

Use of Pb

- Manufacture of battery, cable, pigment, flexible sheet and pipe. Basic Pb carbonate such as $2PbCO_3.Pb(OH)_2$ used in Pb pigment form on basis of grade point. Also, litharge (PbO) used in reverbaratory furnace for oxidation for pigmented varnish and glass production.

- Red Lead (Pb_3O_4): Higher oxide of Pb made by heating Litharge under control condition about 450° C and same use as Litharge.
- Lead Chromate: Normal yellow chromate made by the precipitate from the solution of Lead acitate and nitride to which Sodium Bicarbonate will be added.

Extraction of Ni by Pyrometallurgical Process

Details of Ni

Atomic Number: 28

Atomic Weight: 58.69 amu

Density: 8.98 gm/cm^3

Melting Point: 1452° C

Boiling Point: 2900° C

Common Minerals of Ni

Pentalandite – [$(NiFe)_9S_8$]

Violarate – [Ni_2FeS_4]

Extraction of Ni from Sulphide Ore

Explanation of Flow Sheet

Initial Treatment: The ore is a mixed Cu-Ni ore with nearly equal amount of Cu and Ni. The ore undergoes into grinding and froth floatation to produce a bulk concentrate which sent to copper cliff mill for separation of Cu concentrate, Ni concentrate and Pyrrohotite concentrate with iron sulphide with about 0.8% Ni.

Then the Cu concentrate subjected to O_2 flash smelting for Cu extraction. From Pyrrohotite after roasting iron oxide form and Ni separated by leaching. The Ni concentrate with about 10% Ni, 2% Cu, 40% Fe, and 30% S goes for extraction of Ni in next stage.

Roasting: In conventional process, the concentrate partially roasted to oxidize the iron sulphide either multiple hearth roaster or fluidized bed roaster. But fluidized bed roaster is more preferable because of:

- High output and rich SO_2 gas stream generates.
- Process is Autogeneous.
- Temperature range (550-600)° C about 40% S oxidized.

Smelting: The roasted calcine contains desired amount of siliceous flux is smelted in a reverberatory furnace to produce a matte containing Cu, Ni as 20%, 7% respectively where the slag discarded contains gangue and oxidized iron. There is also converter slag of both Ni and Cu converters are returned to the reverberatory furnace.

Converting: Furnace matte is converted to Ni enriched matte with 50% Ni, 25% Cu, 0.7% Fe, and 21.5% S at 1150° C in Pierce-Smith converter. The slag discarded contains 2% Ni, 1.5% Cu, 40% Fe, and 25% SiO_2 return to reverberatory furnace for recovery of Ni and Cu.

Slow Cooling: Converter matte subjected to slow cooling process from melting point to 400° C for 3 days to form three layers precipitate out as

- First layer, Cu_2S precipitate and grows.
- Second layer, metallic Cu-Ni alloy at 700° C.
- Third layer, solid Ni3S2 phase precipitate at 575° C.

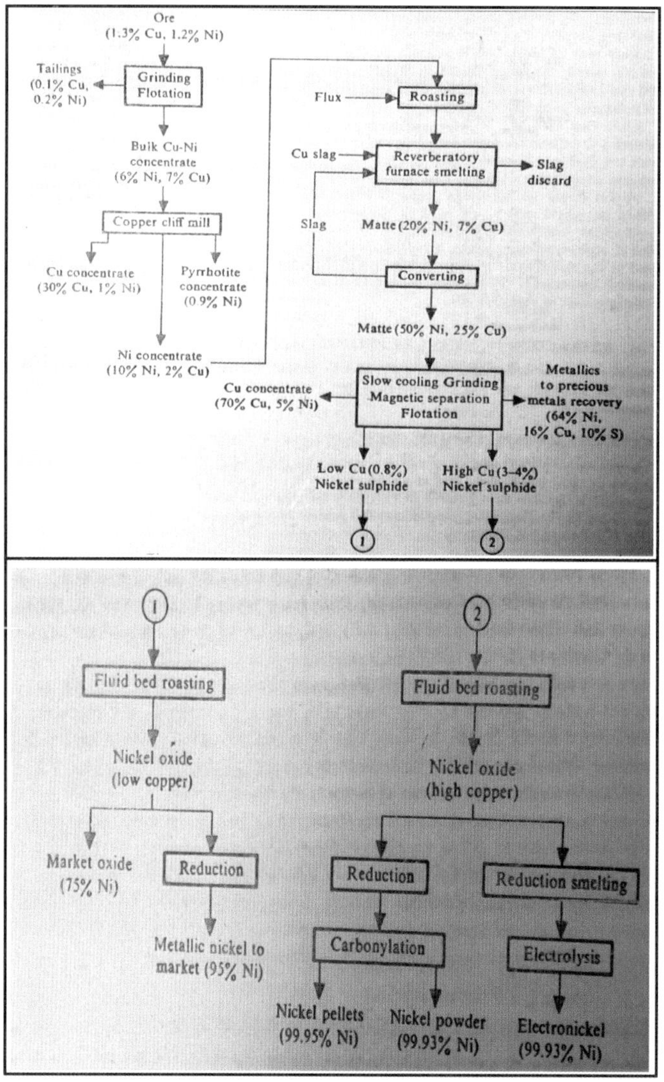

Figure: Nickel Extraction (INCO Process)

Refining Process

Carbonyl Process for Refining Ni

- Mond's Process

In 1889, this refining process of Ni recovered by Carl Langer and Ludwig Mond. In this process, at temperature (40-90)° C metallic Ni combine with CO to give gaseous nickel carbonyl [$Ni(CO)_4$]. At higher temperature (150-300)° C $Ni(CO)_4$ decomposes to give Ni and CO gas. The reactions are

$$Ni + 4CO \xrightarrow{\text{at } 50^0 \text{ C}} Ni(CO)_4$$

$$Ni(CO)_4 \xrightarrow{\text{at } 230^0 \text{ C}} Ni + 4CO$$

Other forms of Carbonyls are volatile carbonyl [$Fe(CO)_5$] Co carbonyl in tetracarbonyl [$Co_2(CO)_8$] tricarbonyl [$Co_4(CO)_{12}$] form. Cu and other major elements are not form carbonyls.

- INCO Process

INCO Atmospheric Carbonylation Process

The oxide first reduces to active Ni in the presence of H2 at about 400° C. Then active Ni undergoes for carbonylation at 50° C to form $Ni(CO)_4$ then at 230° C goes for decomposed to Ni either in pellets about 1 cm dia or powder form about 3.5 μm size.

INCO Pressure Carbonylation Process

The carbonylation reaction has 4 to 1 volume change permits at about 180° C and 70 atm pressure carbonyls of Ni, Fe, and Co formed. From which $Ni(CO)_4$ recovered by fractional distillation and converted to metallic Ni in pellet decomposer or a powder decomposer.

Electrolytic Refining of Ni

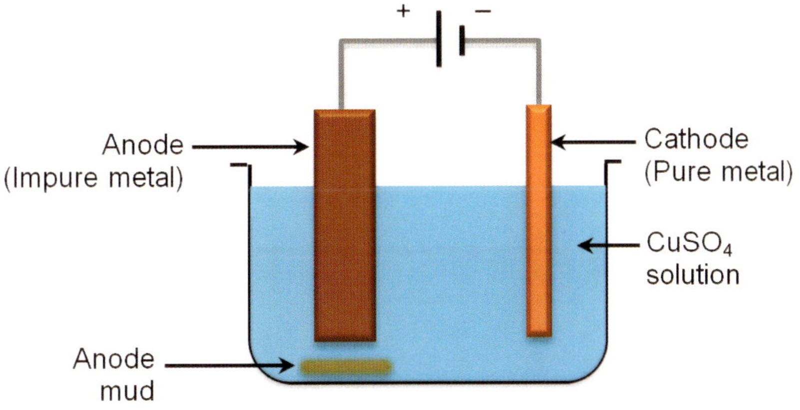

Figure: Electrolytic Refining of Ni

The Ni oxide reduce by coke in fuel fired furnace or electric furnace, and then cast into Ni metal anode. These anodes are electrolytically refined in a bath contains 60 gm/lit Ni^{+2}, 95 gm/lit SO_4^{2-},

35 gm/lit Na^+, 55 gm/lit Cl^-, 16 gm/lit H_3BO_3. This electrolysis carried out at 60° C. Cu remove by cementation with active Ni powder, Fe and other impurities remove by aeration of electrolyte, Co remove by Cobaltic Hydroxide for further Chlorine oxidation. Electrolyzed Ni analyzes about 99.93% Ni.

But in INCO process, electro refining Ni sulphide to metallic Ni to produce anode contains 76% Ni, 0.5% Co, 2.6% Cu, and 20% S. The process is same as the above one, only difference is that the anode enclosed in a bag to collect anode slime contains 95% S and is processed to recover elemental Sulpher and precious metals. The electronickel contains 99.95% Ni.

Uses of Ni

- Important alloying elements used in chemical processing, space research and nuclear reactor.
- Ni-Cu alloys known as Mnel metals are used in chemical oil industry and electric industry.
- Ni, Cu, and Zn alloys are known as German silver. It is used for manufacture of ribbons, bands, and wires for various applications.

Extraction of Magnesium by Pyrometallurgical Process

Details of Mg

Atomic Number: 12

Atomic Weight: 24 amu

Density: 1.738 gm/cm^3

Melting Point: 650 ° C

Boiling Point: 1107° C

Common Minerals of Mg

Dolomite – ($MgCO_3.CaCO_3$)

Magnesite – $MgCO_3$

$MgCl_2$ and $MgSO_4$ present in sea water.

Brucite – $Mg(OH)_2$

Carnallite – $MgCl_2.KCl.6H_2O$

Olivine – Mg_2SiO_4

Sarpentine – $Mg_3Si_2O_7$

Asbestos – $CaSiO_3.2MgSiO_3$

Kainite – $MgSO_4.KCl.3H_2O$

Extraction of Mg

The extraction of Mg occurs by the reaction of MgO with C but not accepted commercially because it is difficult to prevent the backward reaction of Mg vapour with CO.

$$MgO + C = Mg + CO$$

Again MgO reacts with CaC_2 and Al to give Mg as

$$MgO + CaC_2 \rightarrow CaO + Mg + 2C$$

$$3MgO + 2Al \rightarrow Al_2O_3 + 3Mg$$

Again in Pidgeon process MgO reacts with Ferro-Si to give Mg vapour in vacuum as

$$2MgO + FeSi \rightarrow 2Mg + SiO_2 + Fe$$

Dow Process

It is used for production of Mg from sea water containing $MgCl_2$. Sea water used because of

- Economically Mg extract.
- Uniformity of Mg content.

The sea water contains $MgCl_2.1.25\ H_2O$ as feed material. In this process, the following reactions occur as given below

$$CaCO_3 = CaO + CO_2$$

$$CaO + H_2O \rightarrow Ca(OH)_2$$

$$MgCl_2 + Ca(OH)_2 \rightarrow Mg(OH)_2 + CaCl_2$$

$$MgSO_4 + Ca(OH)_2 \rightarrow Mg(OH)_2 + CaSO_4$$

The first step involves collection of Mg as insoluble hydroxide by precipitation with lime according to the above reaction. The suspension is distributed to thickeners, in which hydroxide settles, clear overflow pumped back to sea. The milk of Magnesia underflow is picked up by slurry pumps and delivered to filter tanks. Subsequently; Mg $(OH)_2$ is agitated with 10% HCl. The dilute solution is concentrated and filtered to recover crystallized solids. The filtrate is subjected to evaporation to recover the remaining chloride. During evaporation because hydrous chloride is to lose HCl and result final product is contaminated with oxide. The reaction is

$$MgCl_2.6H_2O \rightarrow MgO + 2HCl + 5H_2O$$

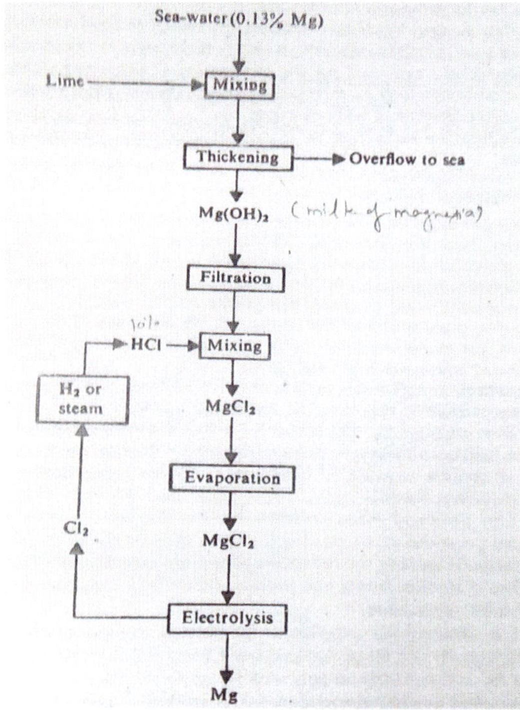

Figure: Production of Mg from Sea Water

Electrolysis of $MgCl_2$

Mg is reactive metal has high decomposition potential. Therefore, aqueous electrolysis H_2 evolution occurs before decomposition of $MgCl_2$. In electrolysis, $MgCl_2$ get fluidized by use of flux and to increase the density of bath so that metal liberated at cathode floats on the surface of electrolyte and not sink.

The electrolytic cells are two types lined and unlined. First one provides separate cathodes made of steel where the later one is steel wall of itself used as cathode. In both cases, graphite used as anode which is inert i. e. not consumed to the metal divert away from anode zone and Chlorine gas collect in a hood and metal to a pool. Chlorine gas from the anode used to dry incoming $MgCl_2$ and chlorinated Mg $(OH)_2$.

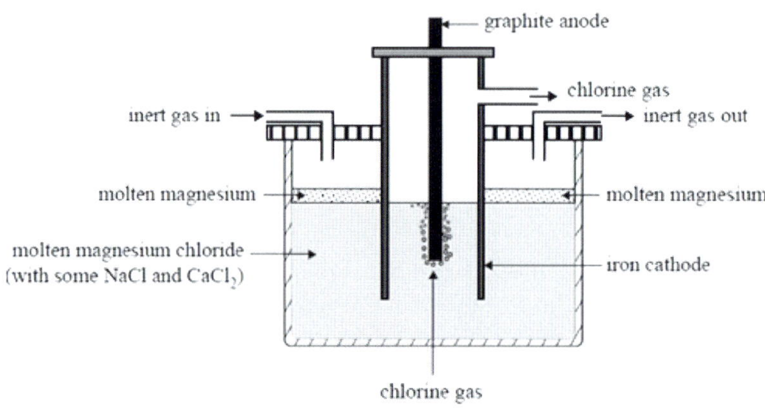

Figure: Electrolytic Cell for Mg Production

Pidgeon Process

Pure Si reduces MgO when both reactants are in standard state. To overcome these difficulties there we will be taken Ferro-Si powder to make in briquette form (75% Si) with calcine dolomite to give Mg in vapour form at (1100-1200)° C under vacuum of 0.1 mm Hg in external of retort as:

$$2MgO.CaO + FeSi \rightarrow 2Mg + 2CaO.SiO_2 + Fe$$

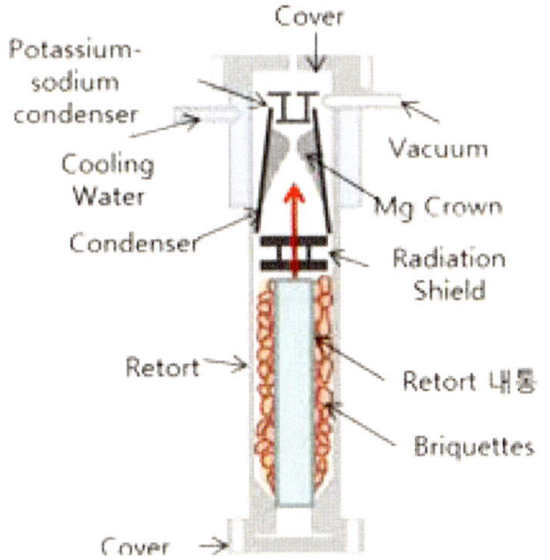

Figure: Retort for Pidgeon Process

The product Mg is gaseous form removed from the retort to avoid possibility of back reaction during cooling and condensation.

The reaction occurs in two ways at 1200° C:

- Absence of lime

  $$2MgO + Si \rightarrow 2Mg + 2SiO_2$$

- Presence of lime

  $$2MgO + 2CaO + Si \rightarrow 2Mg + 2CaO.SiO_2$$

The process is highly endothermic reaction, so large heat input required. There is small addition of CaF_2 or MgF_2 added to speed of the evolution of Mg vapour. Since, vapour pressure of Mg is not altered the effect of fluoride addition is considered catalytic.

Equipment and Operation

The retort is of 25 cm dia. and 3m length inserted horizontally in the furnace made of Nichrome steel with 15% Ni, 28% Cr, 1.25% Mn, 1.5% Si, 0.3% C and Fe balanced in air cooled continuously batch process. The furnace heated either by gas or by electricity.

The charge is of 1:6 ratio of Fe-Si to calcine dolomite of 100 kg weight. Heating carried out to vaporize structurally combined water. The temperature raised about 1150° C and in vacuum. At this stage, Mg vaporizes and condensed by baffles. Application of vacuum stopped and seal broken condensed Mg removed out. Approximately, 120 ton dolomite with 12 tons Ferro silicon gives rise to 10 tons of Mg.

Reaction Mechanism: The reaction occur in following stages:

- Reaction between CaO and Fe-Si produce liquid Ca-Si-Fe. The reaction occur at 1000° C and mildly exothermic. The ternary alloy Ca-Si-Fe serves as a reducing agent.
- Mg vapors produce by reduction of MgO by Ca-Si-Fe alloy. Here, pressure increases and slow down the rate of reaction. Kinetics of the reaction control by the presence of SiO vapour in Pedgeon process as

$$MgO + Si \rightarrow Mg + SiO$$

$$2SiO + 2CaO \rightarrow 2CaO.SiO_2 + Si$$

Magnotherm Process

This process is same as the of Pidgeon process. But the only difference is temperature maintained about 1500° C and the bath maintained in the molten state by addition of alumina to form molten slag.

Uses of Mg

Used of Mg fall under two main heads as

- Non – Structural: Use in alloying, deoxidation and desulphurization of molten metal.
- Structural: Use in automotive industry.

Extraction of Aluminium by Pyrometallurgical Process

Details of Al

Atomic Number: 13

Atomic Weight: 27 amu

Density: 2.702 gm/cm^3

Melting Point: 660.37° C

Boiling Point: 2467° C

Common Minerals of Al

Diaspore – $Al_2O_3.H_2O$ (Al % - 65.4)

Gibbsite - $Al_2O_3.3H_2O$ (Al % - 65.4)

Bauxite – Mixture of Diaspore and Gibbsite

Extraction of Al

Extraction of Al by carbothermic reduction of Al_2O_3 in commercially not adopted because of:

- Required high temperature.
- Undesirable Al carbide form.
- Refractories required for this process are expensive and also, not readily available.

Extraction of Al salt in electrolysis not adopted because of decomposition of aqueous solution results evolution of H_2 gas at lower voltage. So, as Al^{+3} not occur. The H_2 over voltage insufficient to overcome 1.67 volt that the difference in potential between Al deposition and H_2 evolution.

Extraction of Al in Hall – Heroult process: In this process, dissolved Alumina (Al_2O_3) in cryolite ($3NaF.AlF_3$) is electrolyzed, and Al_2O_3 obtain in pure form.

Extraction of Alumina in Bayer's process: In this process, $AlCl_3$ electrolyzed in a fused salt medium contains NaCl and LiCl to obtain alumina (Al_2O_3).

Bayer's Process for Alumina Production

Bauxite initially leached by NaOH under high pressure (25 atm), and temperature (220° C) to form soluble $NaAlO_2$ (Sodium Aluminate) from which $Al(OH)_3$ is precipitate out. When $Al(OH)_3$ calcined, Al_2O_3 or alumina obtained.

Bauxite obtain in mines get crushed and ground to very fine size in jaw crusher and hammer mills, result in high SiO_2 content in finer fractions. Then SiO_2 remove/render by screening in 100 mesh sizes, SiO_2 content is about 10% after communution. Thus leads to enrichment of Al_2O_3 in coarser fraction.

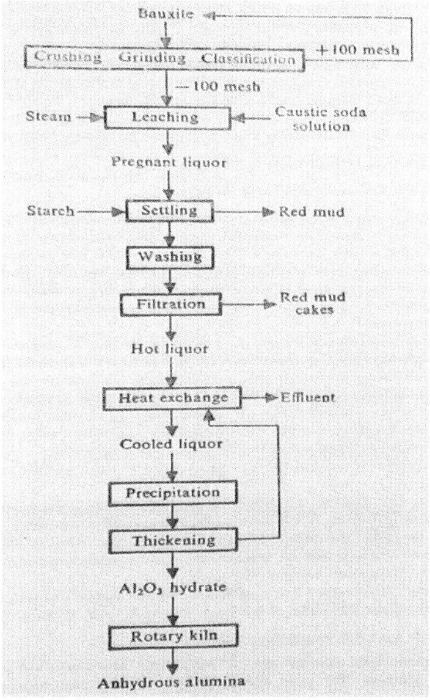

Figure: Flow Sheet of Bayer's Process

Crushed bauxite once again ground with caustic soda (NaOH) in ball mill to yield slurry in which alumina dissolves in bauxite and dissolves in NaOH in a temperature (150- 220)° C and a pressure of (5-25) atm. As a result, impurities leave a suspended solid. The reactions for monohydrate and trihydrate are given below:

$$Al_2O_3.H_2O + NaOH \rightarrow NaAlO_2 + 2H_2O$$

$$Al_2O_3.3H_2O + NaOH \rightarrow NaAlO_2 + 4H_2O$$

Since bauxite is a mixture of mono and trihydrate contains (86-88) % alumina.

Then the liquor cooled little below 100° C and completely pressurized is to settling and clarification section red mud get deposited in presence of starch which speed of the settling. Any remaining red mud gets separated by a series of washers by encountering flow of hot water. After washing the solution the remaining residue consists of ferric hydroxide, silica and alumina.

Hot liquor tapped by heat exchanges to obtain cool liquor. Then cool liquor under precipitation carried out a temperature below critical point. Then $Al(OH)_3$ used for separation precipitate from liquor in a series of thickeners. The product, a coarse hydroxide is finally calcined in a rotary kiln to produce a anhydrous Al_2O_3.

Factors Affecting Bayer Process

- Finer bauxite gives better digestion of alumina in liquid liquor.
- Wet grinding is more efficient than dry grinding.
- Digestion accelerates at high temperature above 100° C.
- At low temperature below a limit cause loss of alumina. So, maintain the temperature at about boiling point of water.
- Sensible heat of recovered from hot liquor by heat exchangers.
- In precipitation stage, entire precipitates of alumina not allow because it may cause precipitation of silica along with alumina.
- For efficient calcinations of alumina in rotary kiln at 1400° C to give anhydrous alumina.

Hall – Heroult Process

This process based on electrolytic decomposition of alumina dissolved in a liquid bath of cryolite. This process requires high purity alumina, cryolite and ash less C electrode. Since alumina is not an ionic. So, it is capable to dissolve in cryolite up to 15% at 1000° C. The synthesize of cryolite occur as reaction of HF with sodium aluminate where, HF form due to the reaction of CaF_2 in $H2_SO_4$ as

$$CaF_2 + (H_2SO_4)_{aq} \rightarrow 2(HF)_{aq} + CaSO_4$$

$$(Na_6Al_2O_9)_{aq} + 12(HF)_{aq} \rightarrow 2Na_3AlF_6 + (6H_2O)_{aq}$$

Cryolite bath get enhanced conductivity by addition of CaF_2 and NaF. At 1000° C density of molten cryolite is 2.1 gm/cm³ and Al_2O_3 is 3.96 gm/cm³. Al liberated heavier than bath and sinks at bottom. Higher the amount of Al_2O_3 in bath, the higher its density and consequently the sinking of Al is rendered more difficult.

Decomposition Potential of Al_2O_3 in dissolved in Cryolite:

- When anode is oxygen, decomposition potential of Al_2O_3 is about 2.1-2.15 v.
- Decomposition potential of graphite at anode is 1.12 v.
- Graphite aids the decomposition of Al_2O_3 because it reacts with O_2 to form CO and CO_2.

$$\frac{1}{2}Al_2O_3+\frac{3}{4}C \rightarrow Al+\frac{3}{4}CO_2$$

$$\frac{1}{2}Al_2O_3+\frac{3}{2}C \rightarrow Al+\frac{3}{2}CO_2$$

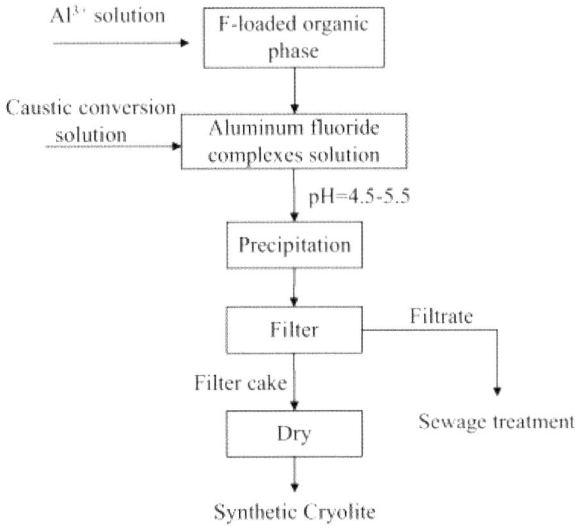

Figure: Synthesis of Cryolite

Influence of H_2 or CH_4 injection at Anode:

$$\frac{1}{2}Al_2O_3+\frac{3}{2}H_2 \rightarrow Al+\frac{3}{2}H_2O$$

$$\frac{1}{2}Al_2O_3+\frac{3}{4}CH_4 \rightarrow Al+\frac{3}{2}H_2+\frac{3}{4}CO_2$$

Injection of H_2 or CH_4 would not be effective in appreciably reducing the decomposition potential of Al_2O_3. Injection may be effectively cut down graphite consumption and prove economical if H_2 or CH_4 available is cheaper. Actual decomposition potential is 5-7 volt, whereas

- Voltage need for electrolytic reduction: 1.7 volt.
- Voltage drop across C lining: 0.6 volt.

- Voltage drop due to anode resistance: 0.5 volt.
- Voltage drop due to resistance of electrolyte: 0.8 volt.
- Voltage drop due to contact resistance: 0.3 volt.

Electrolytic Reduction Cell

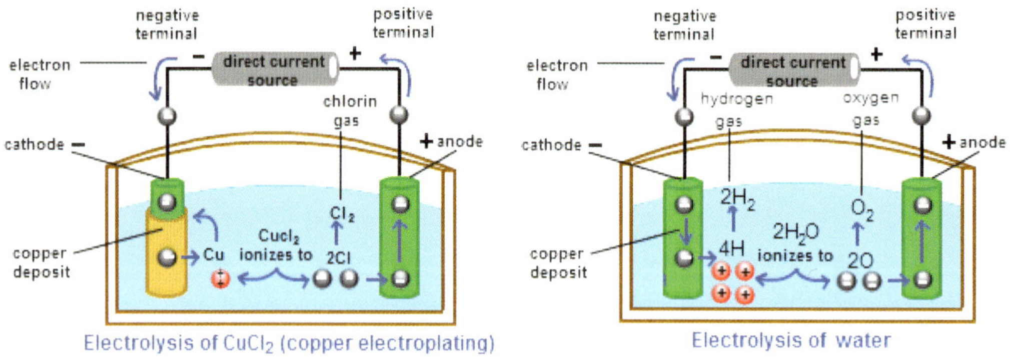

Figure: Electrolytic Reduction Cell

The cell consists of rectangular refractory lining steel box with l=5 m, w=2 m, depth=1 m. Cathode lining consists of refractory bricks faced with C mixed tar binder. A consumable graphite electrode serves as anode. A Fe plate and Fe bar form the cathode are embedded in the C mixture. A tap hole is there, through which metal get collected at the bottom of the hole.

C gets backed in steel mould to produce anode. Fe pins are inserted by electric contact. The baked C electrodes and cathode lining must be sufficiently strong and dense.

Cell Operation

During operation, the cell bath is filled with cryolite and anode lowered into it. A current is passed through it until cryolite melts (M. Pt. – 990° C).

When bath attains a molten state addition of alumina commenced. The alumina decomposes to yield Al and O_2. The necessary heat being supplied by the resistance offered by both the electrodes and the electrolytes.

Anode Effect

If alumina content of bath (normally 5-10%) falls below 2% then normal contact between the anode and bath is interrupted by the gas film in which abruptly increases positive resistance. As a consequence the normal operation ceases. This effect is called anode effect. Under certain conditions, the electrolytes of fused salt is associated with the periodically occurring phenomena in which the terminal voltage increases to high value and current density decreases to a lower value from their normal value. This effect is known as anode effect.

If power consumption, electrolyte, and anode consumption increases during the process. Then, current density exceeds above critical current density.

Factors Affecting Anode Effect

- Nature of electrolyte: Anode effect occurs most readily with fluoride and least with iodides, Bromides and Chlorides required large critical current density than iodide, fluoride both.
- Purity of electrolyte: Greater purity lower in current density and vice-verse.
- Type of electrolytic Anode: The cryolite-Alumina melt anode effect at 4-5 amp/sq. cm. when C anode used and 7-8 amp/sq. cm. for graphite electrode.
- Temperature: Increase in temperature reduces the current density.

Cause of Anode Effect

- Thermal Effect: Local overheating can lead formation of gaseous film which deposits anode from clcctrolytc.
- Electrostatic Causes: During electrolysis, gas bubbles form which is electrically charged. The charge may be originated from the anions discharged at the boundary of newly form gas bubbles at high current density. So, gas is partly ionized when gas bubbles acquires opposite charge to that electrode i. e. negative charge there will be formation of adherent gaseous film which could separate the molten electrolyte from anode.

Metallic Clouds or Fogs

In the electrolysis, fused salt electrolyte, fine globules at the cathode metal get appear in electrolyte is known as metallic clouds.

Al_2O_3 dissolve in cryolite and form Al^{+3}, O^{-2} (in solute) and Na^+, AlF^{3-}, F^- (in solvent).

- Cathodic Reaction:

$$Al_2O_3 \rightarrow Al^{3+} + AlO_3^{2-}$$

$$Al^{3+} + 3\bar{e} \rightarrow Al$$

- Anodic Reaction:

$$2AlO_2^- \rightarrow Al_2O_3 + 3[O] + 2\bar{e}$$

$$6F^- + Al_2O_3 \rightarrow 2AlF_3 + 3[O] + 6\bar{e}$$

Oxygen combines at carbon electrode to form CO_2, according to the reaction as given below

$$\frac{3}{2}CO_2 + 2Al(fog) \rightarrow \frac{1}{2}Al_2O_3 + \frac{3}{2}CO$$

Simple molecular theories of electrolysis are based on the hypothetical dissociation reactions

$$Na_3AlF_6 = NaAlF_4 + 2NaF$$

$$NaAlF_4 = NaF + AlF_6$$

Factors Affecting Electrolysis

- Temperature: Increase in temperature, current efficiency increase. For every 4° C rises 1% current efficiency increase. More number of side reaction and hence dissociation of metal in bath.
- Current Density: Current efficiency increase with increase in current density.
- Inter polar Distance: As the distance increases along the bath from bottom to top, current efficiency also increases.
- Addition of Al_2O_3: At Al_2O_3% is 4%, there no effect on current efficiency. But, decrease or increase in Al_2O_3 from 4% cause increase or decrease in current efficiency.
- Bath Density: With increase in bath density by AlF_3/NaF ratio. But, for large AlF_3% cause bath conductivity and fluidity decreases and volume increases i.e. current efficiency decreases.

Electro Refining of Al

Hoope's Process

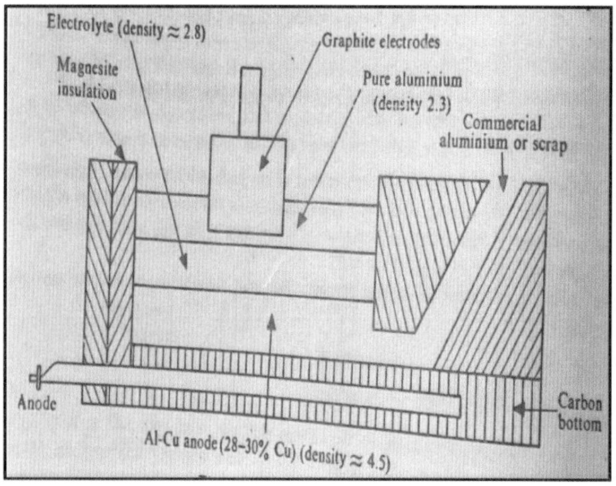

Figure: Three-Layer Process

To achieve high degree of purity three layer processes are employed as

- Bottom or Anode Layer: Specific gravity of this layer is maximum and consists of impure Al alloyed with Cu. This layer serves as anode.
- Middle or Electrolyte Layer: It consists of anhydrous fluorides and chlorides as AlF_3 36%, Barium Fluoride 18%, and CaF_2 16%.
- Top or Cathode Layer: Pure Al layer acts as cathode.

Reaction(at anode): $Al(liq) \rightarrow Al^{3+} + 3\bar{e}$

Reaction(at cathode): $Al^{3+} + 3\bar{e} \rightarrow Al(liq)$

Purity of metal is 99.99%

Alternative Methods of Alumina and Al Production

Alcoa Process

$$Al_2O_3 + 2C + 3Cl_2 \rightarrow 2AlCl_3 + CO + CO_2$$

Al_2O_3 from Bayer process is chlorinated under reducing condition in the presence of C at (700-900)^o C to produce $AlCl_3$, CO, CO_2. $AlCl_3$ vapour form the reaction is condensed at 70° C in a fluid bed containing $AlCl_3$ particles. The solid $AlCl_3$ particles containing feed into an electrolytic cell containing a fused chloride electrolyte ($AlCl_3$+NaCl+LiCl) at 700°C. On electrolysis, Al (liquid) formed at cathode, gaseous Cl_2 liberated at anode.

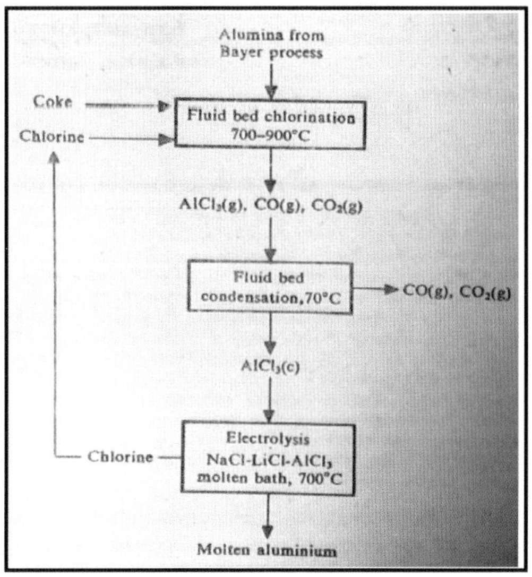

Figure: Flow Sheet of ALCOA Process

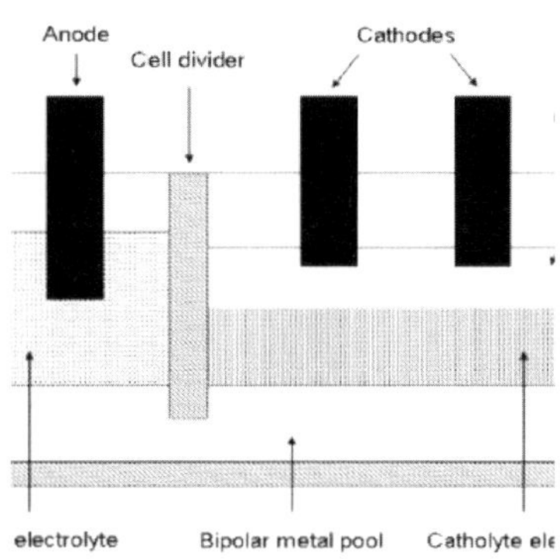

Figure: AL COA Bipolar Cell

Toth Process

A novel process has been proposed by Toth for the extraction of Aluminium. In this process, too, as in the ALCOA process, the alumina from the Bayer process is converted to $AlCl_3$ by chlorination in the presence of carbon. The $AlCl_3$ thus formed is reduced to Al by treating it with Mn at a temperature of 300° C and a pressure of 15 atm. The reaction is

$$8Al_2Cl_3 + 3Mn \rightarrow 3(MnCl_2.2AlCl_3) + 2Al$$

ALCAN Process

- Bauxite reduced to Al alloy.
- Reduction carried out in EAF at 200° C presence of coke.
- The alloy is reacts with preheated $AlCl_3$ vapour in reactor.

$$AlCl_3 + 2Al \xrightarrow{1300^0 C} 3AlCl_3$$

$$3AlCl_3 + 2Al \xrightarrow{700^0 C} AlCl_3 + 2Al$$

Uses of Al

- Use as a good conductor.
- Electrical application.

Extraction of Zirconium

Details of Zr

Atomic Number: 40

Atomic Weight: 91.22 amu

Density: 6.45 gm/cm^3

Melting Point: 1852° C

Boiling Point: 3580° C

Common Minerals of Zr

- Zircon: $ZrSiO_4$
- Mineral associated with beach sands along with other minerals (Ructile, Ilimenite, and Monazite) and oxides of metals such as Fe and Mg.

Extraction of Zr

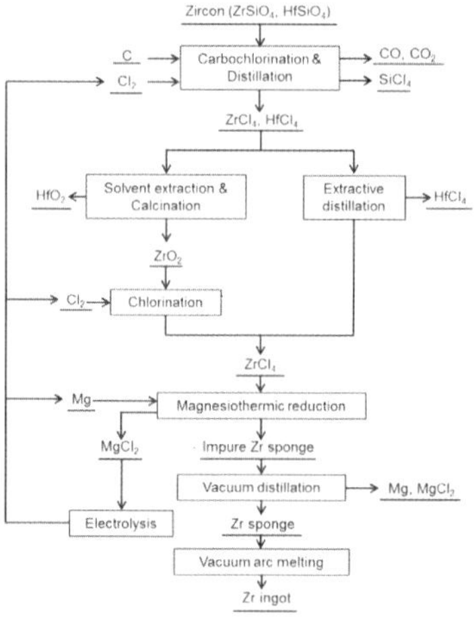

Figure: Production of Zirconium Sponge

As mainly Zr constitute with HF. So our first aim is to separate HF from Zr.

Pyrometallurgical Methods

a. Differential Oxidation of Tetrachloride reaction

$$ZrCl_4 + O_2 \rightarrow ZrO_2 + 2Cl_2$$

$$HfCl_4 + O_2 \rightarrow HfO_2 + 2Cl_2$$

b. Formation of Lower Chloride of Zr reaction as

$$3ZrCl_4 + Zr \rightarrow 4ZrCl_3$$

c. Differential reaction of double chlorides with alkali chloride. The reactions are

$$ZrCl_4 + 2Nacl \rightarrow Na_2ZrCl_6$$

$$HfCl_4 + 2Nacl \rightarrow Na_2HfCl_6$$

Reduction of Zr Compound to Metal

Metallic Zr conventionally produced by metallothermic reduction of Zr tetrachloride. Kroll's process of Mg reduction is most popular method. Recently Zr successfully produced by the fused salt electrolytes of double salts such as Cs_2ZrCl_6 dissolved in alkali chlorides.

Production of Mill Products from Sponge Metal

Conventional induction melting Zr alloys in graphite crucibles in unsuitable due to the prohibitive C pick up. At present, consumable electrode is melting with water cooled Cu moulds employed for tonnage melting of alloys. Procedure evolved for:

- Compaction of sponge briquettes with alloying additions.
- Fabrication of consumable electrode.
- Controlled vacuum arc melting.

For homogenizing the alloy composition and for removing the volatile impurities, double vacuum arc melting is used. The shaping of alloy and extrusion of tubes are achieved through a series of steps, each of which need special precautions.

Uses of Zr

- Non nuclear applications: Flash bulbs in the form of thin foils and ceramics as an oxide.
- Nuclear reactor used as cladding element.

Extraction of Titanium

Details of Ti

Atomic Number: 22

Atomic Weight: 47.90 amu

Density: 4.54 gm/cm^3

Melting Point: 1668° C

Boiling Point: 3260° C

Common Minerals of Ti

Rutile: TiO_2

Ilimenite: FeO. TiO_2

Extraction of Ti

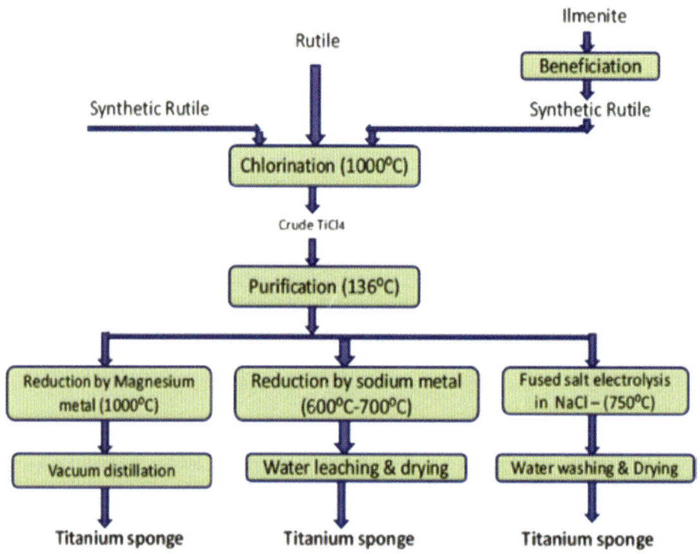

Figure: Production of Graded Titanium Sponge

Kroll's Process

The reduction reaction in Kroll's process can be written as

$$TiCl_4 + 2Mg \xrightarrow{800^0 C} Ti + 2MgCl_2$$

The reaction is carried out in a stainless steel container for externally heating by gas or electricity. Exothermic reaction occurs. After reaction is over the Ti sponge is recover either by dissolving $MgCl_2$ and excess Mg by leaching with cold dilute hydrochloric acid or by distilling off $MgCl_2$ and excess Mg by vacuum heating. The residue is pure Ti sponge. The maximum yield is 96% because of:

- Efficient Scavenger w.r.t impurity chlorides such as $FeCl_3$ and $AlCl_3$.
- Better conductivity of NaCl compares to $MgCl_2$ and hence, reaction is exothermic.
- NaCl-$MgCl_2$ leads a better separation of metal from slag.

Hunter Process

The reduction of $TiCl_4$ by Na is to give Ti known as Hunter process and written as

$$TiCl_4 + 2Na = Ti + 4NaCl$$

The reaction is highly exothermic, there is a series of reactions occur as

$$TiCl_4 + Na = TiCl_3 + NaCl$$

$$TiCl_3 + Na = TiCl_2 + NaCl$$

$$2TiCl_2 = Ti + TiCl_4$$

$$2TiCl_3 = TiCl_2 + TiCl_4$$

$$TiCl_2 + 2Na = Ti + 2NaCl$$

$$3TiCl_2 = Ti + 2TiCl_3$$

$$TiCl_4 + 2Na \rightarrow TiCl_2 + 2NaCl$$

$$4TiCl_3 = Ti + 3TiCl_4$$

$$TiCl_3 + 3Na = Ti + 3NaCl$$

The reaction of sodium with titanium subchlorides in fused sodium chloride is an electro chemical nature and occurs same as like galvanic corrosion of iron.

Reaction(anode): $2Na(sol) = 2Na^{+}(sol) + 2\bar{e}$

Reaction(cathode): $Ti^{+}(sol) + 2\bar{e} = Ti(c)$

The anode reaction takes place at a metal surface where sodium metal, fused NaCl, Cl^{-} ions are available for salvation of Na^{+} ions. The Cathodic reaction takes place at a metal site where soluble Ti in fused salt is available and Cl^{-} ions released completely in the anodic reaction.

Uses of Ti

The sponge Ti produces in Kroll's process and ductile Ti produce in Hunter process. Main uses of Ti overally is:

- Jet engine components.
- Air frames.
- Missiles and space craft.

Extraction of Tin (Sn)

Details of Tin

Atomic Number: 50

Atomic Weight: 118.71 amu

Density: 7.31 gm/cm^3

Melting Point: 231.9^o C

Boiling Point: 2270°C

Common Minerals of Tin

Cassiterite: SnO_2

Extraction of Tin

The concentration contains FeS are oxidize to oxide during smelting. Carbothermic reduction of SnO_2 Extraction of Tin: 2 is feasible at moderate temperature. At higher temperature (1200-1300)^o C FeO is more stable. The difference in stability between FeO and SnO_2 used as the basis of selectivity of reducing SnO_2 to Sn. The difference in stability further reduce because slag has higher affinity for Tin oxide compare with iron oxide and tendency for attain separation of Tin from Fe becomes difficult. So smelting of Tin has done in 3 ways.

First stage, 99% obtain partially reduction of tin concentration almost all Fe in slag. The slag further reduced in 2 stages. First stage produce tin contain 5% and in the second stage tin contain 20% Fe. Third stage have very low % of tin. So, 2nd and 3rd stage subjected to liquation to eliminate Fe.

When Sn-Fe alloys heated to a temperature higher than melting point of Sn. The metal known as first run metal contain 99.9% Sn. Temperature progressively increase Fe content in Sn increase known as second run metal. Where, for taking smelting in rotary instead of reverberatory furnace of higher efficiency Sn recovery. First smelting slag is reduce and Sn volatile either in form of SnO or SnS. Clean slag discarded after reduction vapour pressure of SnS gteater than SnO. So, SO_2 presence is beneficial.

Refining of Tin

There is mainly two methods of refining of Sn as Pyrometallurgical refining (or fire refining), and electro refining.

Pyrometallurgical Refining: Ultimate recovery of Tin is more than 85%. Principle is to several common impurities from intermetallic compounds with each other or with Sn. Thus, if impure liquid tin is cooled to a temperature just above the melting point of tin (232° C), these inter metallic compounds as Cu_2Sb, Cu_3Sn, Cu_3As, $FeSb_2$, $FeSn_2$, FeAl etc. have higher melting point than tin. So separate out and get floats. S adds to remove Cu in CuS form. AlAs and AlSb by addition of Al, As, and Sb eliminated.

$$2AlAs + 6H_2O = 2AsH_3 + 2Al(OH)_3$$

$$2AlSb + 6H_2O = 2SbH_3 + 2Al(OH)_3$$

NH_4Cl add with tin after Al treatment to remove surplus Al as

$$Al + 3NH_4Cl = AlCl_3 + 3NH_3 + \frac{3}{2}H_2$$

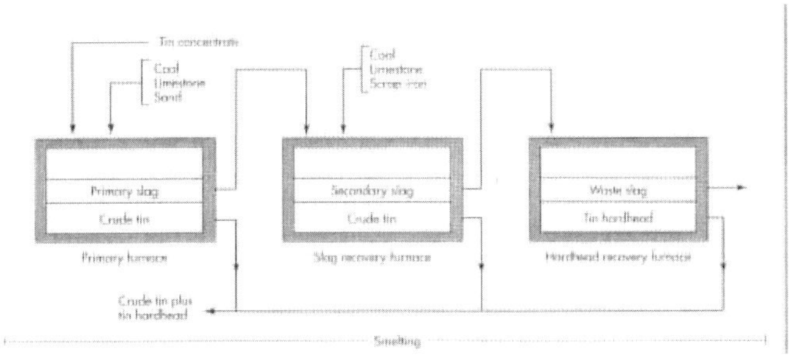

Figure: Smelting of Tin Concentrate

Pulverize coal separate and precipitate out $AlCl_3$. Deleading of Sn by $SnCl_2$ based on

$$Pb + SnCl_2 = PbCl_2 + Sn$$

A Sn-Pb alloy removes from $PbCl_2$-$SnCl_2$ mixture by melting in contact with Zn.

$$SnCl_2 + PbCl_2 + 2Zn = 2ZnCl_2 + Pb - Sn$$

Any excess Ca or Mg is removed by adding NH_4Cl according to the reaction

$$2NH_4Cl + Ca(Mg) = 2NH_3 + H_2 + Ca(Mg)Cl_2$$

Electro Refining of Sn: Using a mixture of H_2SO_4 and phenol sulphonic acid as electrolytes is feasible. If, Sn contains large amount of impurities such as Pb, Bi, Sb, As etc. Mixture of Cresol sulphonic acid and phenol sulphonic acid now used as electrolyte.

It is analyze 95% Sn dissolve in pure metal to obtain at cathode sheet of 99.95% Sn.

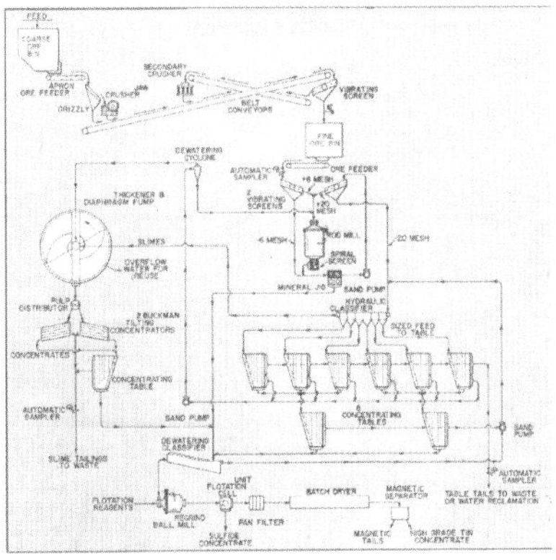

Figure: Flow Sheet of Production of Refined Tin

Use of Sn

- Resistance to both corrosion and toxicity. So, for plating cans, general plating other materials.

Extraction of Nb (Colombium)

Details of Nb

Atomic Number: 41

Atomic Weight: 92.91 amu

Density: 8.57 gm/cm³

Melting Point: 2750 K

Boiling Point: 5017 K

Common Minerals of Nb

Columbite – (Fe, Mn)(Nb, Ta)$_2$-O$_6$

Pyrochlore – (Ca, Na)$_2$(Nb, Ta, Ti)$_2$O$_6$(OH,F)

The extraction of Nb follows the same extraction route as Ta. The figure represents the extraction route of Nb and Ta.

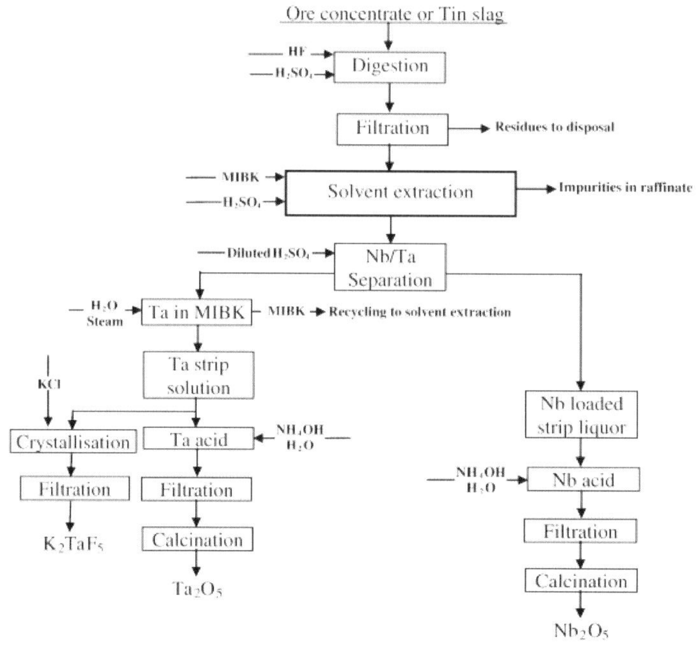

Figure: Extraction of Niobium and Tantalum.

Use of Nb

- Niobium is used in alloys including stainless steel. It improves the strength of the alloys, particularly at low temperatures. Alloys containing niobium are used in jet engines and rockets, beams and girders for buildings and oil rigs, and oil and gas pipelines.

- This element also has superconducting properties. It is used in superconducting magnets for particle accelerators, MRI scanners and NMR equipment.
- Niobium oxide compounds are added to glass to increase the refractive index, which allows corrective glasses to be made with thinner lenses.

Extraction of U (Uranium)

Details of Uranium (U)

Atomic Number: 92

Atomic Weight: 238.03 amu

Density: 19.1 gm/cm^3

Melting Point: 1405.3 K

Boiling Point: 4404 K

Common Minerals of Uranium (U)

Uraninite – UO_2

Extraction of Uranium (U)

The extraction route of Uranium (U) follows the flow sheet given in figure.

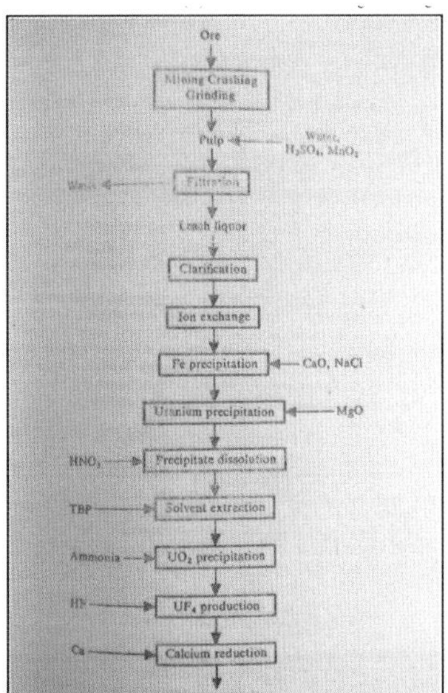

Figure: Extraction of Uranium Ingot

Use of Uranium (U)

- Uranium is a very important element because it provides us with nuclear fuel used to generate electricity in nuclear power stations. It is also the major material from which other synthetic transuranium elements are made.

- Naturally occurring uranium consists of 99% uranium-238 and 1% uranium-235. Uranium-235 is the only naturally occurring fissionable fuel (a fuel that can sustain a chain reaction). Uranium fuel used in nuclear reactors is enriched with uranium-235. The chain reaction is carefully controlled using neutron-absorbing materials. The heat generated by the fuel is used to create steam to turn turbines and generate electrical power.
- In a breeder reactor uranium-238 captures neutrons and undergoes negative beta decay to become plutonium-239. This synthetic, fissionable element can also sustain a chain reaction.
- Uranium is also used by the military to power nuclear submarines and in nuclear weapons.
- Depleted uranium is uranium that has much less uranium-235 than natural uranium. It is considerably less radioactive than natural uranium. It is a dense metal that can be used as ballast for ships and counterweights for aircraft. It is also used in ammunition and armour.

Ferrous Metallurgy

Ferrous metals are those metals that are primarily composed of iron and have magnetic properties in them. Some common ferrous metals include alloy steel, carbon steel, cast iron and wrought iron. These metals are easily recycled. Carbon steel is a common ferrous metal that is widely recycled and reused.

Ferrous metals are known for their hardness, durability and tensile strength. Ferrous metals are widely used in almost all industries such as the manufacture of shipping containers, industrial piping, automobiles, railroad tracks, and many commercial and domestic tools.

Ferrous metals are highly vulnerable to rusting and corrosion when exposed to moisture or an acidic or corrosive environment due to the high content of carbon molecules. However, this is not true with wrought iron; it resists rust due to its purity. Stainless steel is protected from rust by the presence of chromium.

Most ferrous metals have good magnetic properties and are considered to be good conductors of electricity. This property makes them suitable for electronics.

Commonly used ferrous metals are:

- Steel
- Carbon steel
- Alloy steel
- Cast iron
- Wrought iron

The early application of iron to the manufacture of tools and weapons was possible because of the wide distribution of iron ores and the ease with which iron compounds in the ores could be reduced by carbon. For a long time, charcoal was the form of carbon used in the reduction process. The production and use of iron became much more widespread about 1620, when coke was introduced as the reducing agent. Coke is a form of carbon formed by heating coal in the absence of air to remove impurities.

The overall reaction for the production of iron in a blast furnace is as follows:

$$Fe_2O_3(s) + 3C(s) \xrightarrow{\Delta} 2Fe(l) + 3CO(g)$$

The actual reductant is CO, which reduces Fe_2O_3 to give Fe(l) and $CO_2(g)$ ($Fe_2O_3(s) + 3C(s) \xrightarrow{\Delta} 2Fe(l) + 3CO(g)$; the CO_2 is then reduced back to CO by reaction with excess carbon. As the ore, lime, and coke drop into the furnace, any silicate minerals in the ore react with the lime to produce a low-melting mixture of calcium silicates called slag, which floats on top of the molten iron. Molten iron is then allowed to run out the bottom of the furnace, leaving the slag behind. Originally, the iron was collected in pools called pigs, which is the origin of the name pig iron.

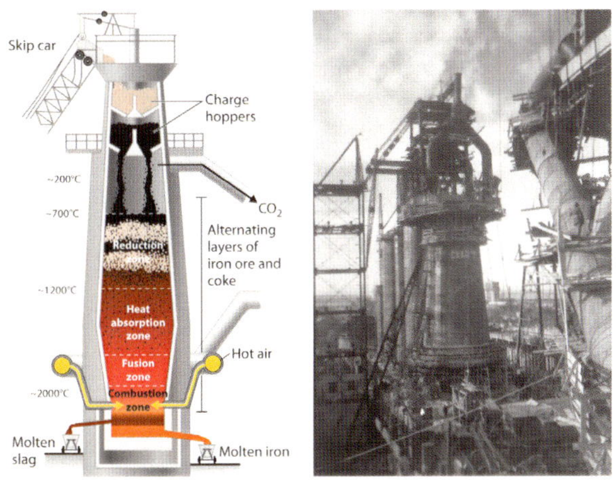

(a) Blast furnace (b) World's largest blast furnace in 1931

A Blast Furnace for Converting Iron Oxides to Iron Metal.

(a) The furnace is charged with alternating layers of iron ore (largely Fe_2O_3) and a mixture of coke (C) and limestone ($CaCO_3$). Blasting hot air into the mixture from the bottom causes it to ignite, producing CO and raising the temperature of the lower part of the blast furnace to about 2000° C. As the CO that is formed initially rises, it reduces Fe_2O_3 to form CO_2 and elemental iron, which absorbs heat and melts as it falls into the hottest part of the furnace. Decomposition of $CaCO_3$ at high temperatures produces CaO (lime) and additional CO_2, which reacts with excess coke to form more CO. (b) This blast furnace in Magnitogorsk, Russia, was the largest in the world when it was built in 1931.

The first step in the metallurgy of iron is usually roasting the ore (heating the ore in air) to remove water, decomposing carbonates into oxides, and converting sulfides into oxides. The oxides are then reduced in a blast furnace that is 80–100 feet high and about 25 feet in diameter in which the roasted ore, coke, and limestone (impure $CaCO_3$) are introduced continuously into the top. Molten iron and slag are withdrawn at the bottom. The entire stock in a furnace may weigh several hundred tons.

Within a blast furnace, different reactions occur in different temperature zones. Carbon monoxide is generated in the hotter bottom regions and rises upward to reduce the iron oxides to pure iron through a series of reactions that take place in the upper regions.

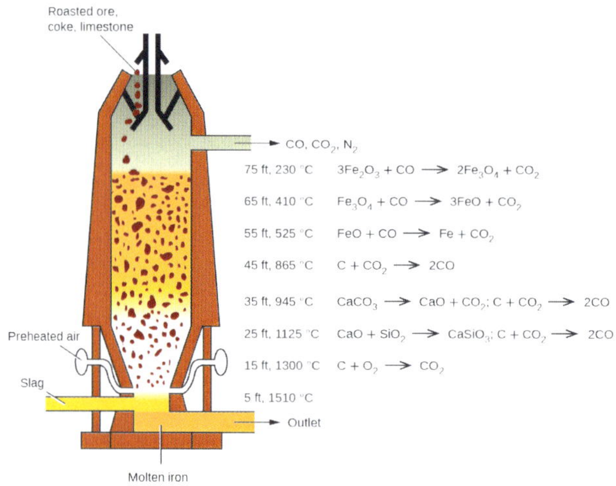

Near the bottom of a furnace are nozzles through which preheated air is blown into the furnace. As soon as the air enters, the coke in the region of the nozzles is oxidized to carbon dioxide with the liberation of a great deal of heat. The hot carbon dioxide passes upward through the overlying layer of white-hot coke, where it is reduced to carbon monoxide:

$$CO_2(g) + C(s) \rightarrow 2CO(g)$$

The carbon monoxide serves as the reducing agent in the upper regions of the furnace. The iron oxides are reduced in the upper region of the furnace. In the middle region, limestone (calcium carbonate) decomposes, and the resulting calcium oxide combines with silica and silicates in the ore to form slag. The slag is mostly calcium silicate and contains most of the commercially unimportant components of the ore:

$$CaO(s) + SiO_2(s) \rightarrow CaSiO_3(l)$$

Just below the middle of the furnace, the temperature is high enough to melt both the iron and the slag. They collect in layers at the bottom of the furnace; the less dense slag floats on the iron and protects it from oxidation. Several times a day, the slag and molten iron are withdrawn from the furnace. The iron is transferred to casting machines or to a steelmaking plant.

Molten iron is shown being cast as steel.

the other is heterogeneous. In the most of the alloys, heterogeneous transformation is observed. The rate of heterogeneous is determined by the rate of the nucleation and growth. The rate of nuclei is expressed in terms of number of nuclei formed per unit time and per unit volume and the growth rate is expressed in terms of cm per second.

4. Carburizing: This is the process of measuring the diffusion coefficient of carbon in steel.
5. Hardenability of steel: The susceptibility to hardness of steel is measured by quenching. The susceptibility of steel shows the depth of hardening of steel under given cooling condition. The hardenability is related to the isothermal transformation. Steel will harden up to high degree with low critical cooling rate than steel with high critical cooling rate.
6. Eutactoid transformation in steel: This involves the measurement of inter-lamellar spacing and volume fraction of pearlite. When the high temperature of austenite transforms into the aggregate of two layers ferrite and cementite is known as pearlite. This transformation is affected by time, temperature and pearlier morphology factors.
7. Tempering of mertencite: It involves the study of decomposition of mercentite in the process of tempering. Tempering is the simple heat treatment which is used to increase the physical properties of quenched steel.
8. Electrical resistivity of alloy: It includes the calculation of band gap energy in the semiconductors like germanium etc.
9. Electrical conductivity of ionic solid: The electrical conductivity of ionic solids is due to the migration of ions in the influence of electric field. The migration is because of presence of some lattice defects in crystal.
10. Dielectric behavior of barium titanate: The dielectric behavior of barium titanate is measured by calculate the dielectric constant and Curei-weiss temperature. The polymorphic transformation occurs in barium titanate. It is from Rhombohedral to Orthorhombic to Tetragonal to Cubic. All these are interconvertible to each other at particular temperature.

Chemical Metallurgy

Chemical metallurgy is the branch of Metallurgy that deals with the extraction of metals from naturally occurring compounds and their refinement to levels of purity suitable for commercial use. The starting raw materials, however, in many cases are not restricted to naturally occurring compounds. Nor in all cases are the end products desired as metals. The chemical treatments quite often depend upon the prior physical processing to which the relevant materials have been subjected. In acknowledgement of these facts it is only proper that both physical and chemical treatments are considered together in the total scheme defined for materials processing. There exist a very large variety and range of treatments and many possible combinations of these which may be used in chemical metallurgy processes. The chemical metallurgy coverage extends further to deal with all the chemical properties of metals, including those pertaining to the formation of alloys by uniting one metal with other metals or non-metals.

Physical metallurgy has close links with chemical metallurgy. It helps in obtaining a scientific understanding of the various chemical metallurgy processes and can assist in designing new processes or modifications of existing processes. The linkage between physical metallurgy and chemical metallurgy has grown with time. This has helped considerably in replacing the empirical or the "trial and error" mode of development with more logical and science based modes. For example, a newly found ore resource can be studied by such powerful physical metallurgy tools as x-ray diffraction, electron microscopy and electron probe microanalysis, and with the knowledge thus gained about the resource, it is possible to make very useful predictions in a number of fronts. Once the type and the nature of the minerals present in a given resource are studied and analysed, it is entirely possible to make useful predictions on the processing of the resource. Refractory minerals would need highly corrosive leachants and aggressive conditions of leaching such as those obtained at elevated temperatures and pressures as in autoclaves. If the structural features show the substitution of one element at the lattice sites of another then separating these elements by physical methods is not possible.

The area of mineral dressing concerns itself with the starting operations and involves crushing and grinding either for purposes of liberation of the desired substances that are physically entrapped with others or for making them adoptable for further processes. Once liberated, they normally go through such physical beneficiation processes as magnetic separation, electrostatic separation, screening, flotation, to name some, to produce a beneficiated product usually called a concentrate and chemical metallurgy normally takes over from then onwards.

The developments in chemical metallurgy pertain to these improvements. Much development can be expected in the coming decades in chemical metallurgy on account of the efforts needed to extract purer metals from leaner and more complex ores with minimum energy expenditure, while causing minimum environmental pollution. These demands will keep research on improved and new extraction processes alive and make extractive metallurgy an important and dynamic branch of metallurgy, as in the past.

Pyrometallurgy, the dominant process in chemical metallurgy, uses reactor of different types and designs. In terms of the physical states of the reactants, one generally finds that the different reactions carried out in pyrometallurgy include principally, gas/liquid, liquid/ liquid, gas/solid, and solid/solid reactions.

Table: Reactors for agitation leaching

Reactor name	Description
Pachuca tank	The simplest and most inexpensive device for agitating the pulp. Agitation is achieved by the injection of air under pressure into the bottom of a tall tank. The rising gas draws the surrounding liquid with it to the surface.
Low-speed agitators	These are suitable for systems not needing high turbulence. Large diameter paddles or rakes may be used to mix slurries containing fine material.
High-speed agitators	These are suitable for leaching reactions that are limited by diffusion in the leach liquor. An agitator of this type requires a propeller or a turbine to provide high-speed agitation.
Agitated autoclaves	These systems are used for leaching under conditions of elevated temperatures and pressures.

These reactions are implemented in different types of reactors. The classification of reactors or furnaces may be: batch type, continuous type, direct heating type, and indirect heating type. Some of the furnaces used in the main metallurgical operations are: crucible furnace; hearth furnace-open hearth; shaft furnace blast furnace; converters–Bessemer, side-blown or top-blown (LD); muffle furnace; soaking pits; reheating furnaces; and annealing furnace. Among the other types of furnaces mention may be made of rotary kiln-cement kiln; tunnel kiln; boiler furnace; electric furnaces (three-phase or single-phase, heating done by metallic resistors, or by induction). It may, however, be pointed out that the classification of furnaces does not serve such a useful purpose as classification of the processes that occur in the furnace. Two furnaces may be alike as far as one process is concerned, but entirely dissimilar from the viewpoint of another process.

It is clear that many of the chemical metallurgy processes must be carried out at high temperatures. In this respect, it is necessary to be acquainted not only with the process reactors but also with the methods of heat generation and with the refractories that are needed in the reactors to cope up with the high temperatures attained.

Electrometallurgy

Electrometallurgy deals with the use of electrical energy for production of pure metals by electrolysis. Electrometallurgy is usually the last stage in metal production after corresponding pyro- or hydrometallurgical operations. The process within the metal is produced by its aqueous salt as a last stage of the hydrometallurgical operations, is known as electrowining. Electrolytic purification of the crude metal produced by pyrometallurgical operations is known as electrorefination. Both processes are based on electrolysis, i.e. on the process in which forced electrons exchange of the matter occurs. Electrolysis, in some cases, is an irreplaceable way for production or refination of some metals. For example, aluminum cannot be produced by pyrometallurgical procedure. Electrorefination of metals is unique process considering the level of metal purity, energy consumption as well as possibility for extraction of precious metal.

Electrolysis Cell

Basic consisting elements of the electrolytic cell are shown in the simplified scheme in figure. Two electronic conductors – electrodes (anode and cathode), an ionic conductor (electrolyte) and an energy source (battery) are necessary to perform electrolysis. In the electrolysis cell electrode reactions are non-spontaneous, i.e. they are forced by an external source of energy. So, reduction of the metal ions to pure metal (production or refination of metal) is enabled. For efficient electrolysis, the above elements have to satisfy some criteria. So, electrodes have to:

- Be made by an appropriate material that promotes corresponding electrode reactions and exhibits good long-term stability and is environmental friendly;
- Be appropriately processed in order to remove electrode products easily. Cathodic edges should be rounded off to prevent dendritic growth;
- Be appropriately arranged into the cell for better volume efficiency. The shorter cath-

ode-anode distance, the lower the Ohmic drop in the cell and consequently the lower energy consumption. On the other side, grown deposit on the cathode may cause short cut if the electrodes are too close.

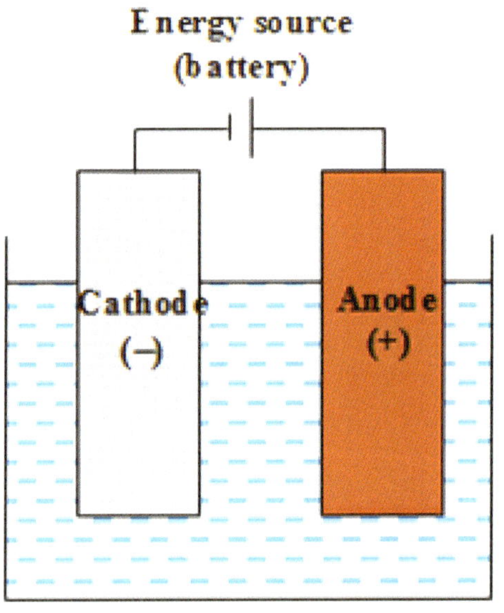

Figure: Simplified scheme of the electrochemical cell

The electrolyte should contain:

- Compound of the ion – participant in the electrode reaction;
- Compounds that provide optimal electrical conductivity, corresponding acidity/ alkality and/or complex forming ability;
- Corresponding additives for regulation of the surface tension, to optimize the diffusion potential, to provide better adhesion of the deposit over the cathodic substrate, to reach better optical and mechanical properties of the deposit;
- In some cases – mechanical components, e.g. thin oil layer or inert plastic balls in order to reduce electrolyte evaporation;
- Additives for optimizing density and viscosity in the case of the molten salt electrolysis.

The source of energy should be:

- Properly designed to provide voltage and current output high enough for performing electrolysis;
- Equipped with voltage regulation system;
- Long-term stable.

The electrolysis cell has to be:

- Prepared by a material having good chemical, thermal and mechanical stability. Traditional materials are wood, glass, concrete, stainless metals/alloys, while the recent ones are various polymer materials or coatings;

- Equipped to shoulder electrodes (whose weight can reach even tens of tones) and to supply them with electricity;
- Equipped for heating/cooling, stirring/ circulation and filling/emptying the electrolyte;

Electrorefination vs. Electrowining

Depending on the input materials in the cell, electrolysis is aimed to produce or refine the metals. In the case of electrorefination, anodes are prepared by a raw metal produced by pyrometallurgical process. So, the basic metal ions originate by dissolving of the anode and further, they reduce on the cathode where the final product is highly-pure metal. In this case, there are soluble anodes which are the main source of basic metal ions, while the electrolyte possesses basic metal ions in quantity only to maintain sufficient concentration to avoid deposition of impurities on the cathode. In the case of electrowining the source of basic metal ions is the electrolyte. The anode docs not dissolve during this process. On the anode some other anodic reaction occurs, e.g. oxygen or chlorine evolution. For this reason they are called non-soluble anodes.

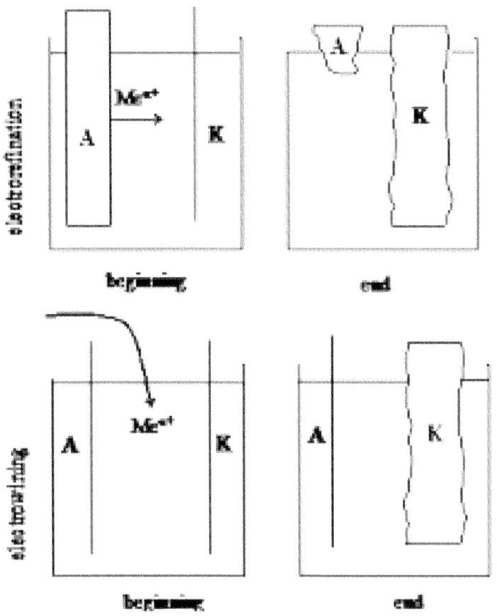

Simplified schemes of electrorefination and electrowining processes

In the figure simple schemes of electrorefination and electrowining processes are shown for better understanding of the differences between them. In both cases cathodic process is metal ions reduction, i.e. electrodeposition of the metal. In order to produce polycrystalline and compact deposits, it is important to know the fundamentals of the electrocrystallization and the correlation between electrochemical parameters (overpotential, current density) and metal morphology. We address the readers to corresponding literature on electrochemical deposition of metals. To produce purer and more compact metallic product, it is necessary to provide:

- Better quality and arrangement of the electrodes. The purest deposit can be produced if electrodes are arranged vertically and maintain constant optimal distance between anode and cathode;
- Optimal composition of the electrolyte to maintain both continuous constant supplying the cathodes with basic metal ions and low concentration of impurity metal ions. In the case

of electrorefining, purification of the electrolyte should be performed in order to reduce the concentration of impurities and to avoid their deposition on the cathodic product. The concentration of the acid (mostly sulfuric acid) should be high enough to provide high electrical conductivity, i.e. lower Ohmic drop of the electrolyte. In order to smooth the metal deposit and to control the crystal growth, corresponding additives are used, e.g. gelatin as deposit smoother or thiourea and chloride ions as crystal growth regulators;

- Optimal temperature. Heating the electrolyte causes higher electric conductivity and consequently lower energy consumption. Also, the solubility of the basic metal ions is higher and consequently their diffusion to the cathode is made easier, as well as the rate of the charge transfer on the cathode is higher. At higher temperatures electrolyte density and viscosity decrease that is suitable for avoiding particles of anode mud to be included into the cathodic product. But, the working temperature should have optimal value (not so high) due to evaporation of the electrolyte and higher energy consumption. In the case of electrowining of electronegative metals such as Zn, In or Cd, overpotential of hydrogen evolution on the cathode is made easier. The presence of hydrogen is very harmful for the cathodic product;
- Optimal current density. At higher current density, electrodeposition is faster (desirable), but the dendritic growth is forced (undesirable). Dendritic deposits enslave electrolyte and anode mud particles more easily. Also, dendritic structure of metals is technically nonapplicable.

Aqueous vs. Molten Salts Electrolysis

Electrolysis in aqueous electrolytes is a suitable process for metal production and refination at low temperature. As a result, cost-effective metal of high purity can be produced. The main solvent of the electrolytes for electrolysis is the water. There is not other solvent in electrochemistry that possesses unique water properties, such as:

- Polarity – prerequisite for dissolving ionic compounds,
- Electrolytic dissociation – prerequisite for better electric (ionic) conductivity of the electrolyte,
- Stability (physical, chemical and thermal),
- Wide interval of liquid state existence,
- Non-toxic, non-corrosive and non-complexable, – availability and – inexpensive.

But, electrolysis in aqueous electrolytes is limited to metals more positive than hydrogen such as Cu, Ag, Au, and more negative ones that have high over potential for hydrogen evolution such as Zn, Cd, In etc. The apparent electronegative metals such as Na, Mg, Al, K, Li cannot be produced by electrolysis in aqueous solution because on the cathode only hydrogen evolution occurs. In order to reduce these metals on the cathode, $H+$ ions have to be removed from the electrolyte, i.e. have to find other solvent which have above characteristics. Unfortunately, there is not some other inorganic or organic solvent with corresponding properties in industrial practice. So, the other approach to solve this problem is using molten metal salts. This is still a unique way for production of alkaline and alkaline earth metals (Na, K, Mg) and Al. Their oxides are very stable and their pyrometallurgical production is technically and commercially non-effective. Molten salt electrolysis has more disadvantages related to the aqueous electrolysis, such as:

- Considerably higher working temperatures (up to 900 o C in case of Al),
- Complex composition of the electrolytes,
- Highly-aggressive electrolytes,
- Complex electrode reactions,
- Low current efficiency, i.e. high specific energy consumption and
- Anodic effect – Stop of the current pass along with high rise of overvoltage, which causes breaking off of the electrolysis process.

Economic / Energetic Aspects

Electrometallurgy is one of the highest consumers of energy within the global energy consumption. For illustration, two thirds of entire produced energy is consumed by the industry, while within the industry consumption, two thirds are consumed by the electrometallurgy. In table data about the energy consumption for production of the most important technical metals are shown.

In the time of enormous energy consumption, as well as the soon exhausting of the most used energent – fossil fuels, the present imperative of the electrometallurgy is the decrease of the energy consumption. To save energy consumption, it is important to know the specific energy consumption of electrolysis and structure of the applied voltage. Specific energy consumption (w) is defined as a ratio of the energy consumption of the process (W) and corresponding produced matter (m):

$$w = \frac{W}{m} = \frac{U \cdot I \cdot \tau}{I \cdot \tau \cdot q \cdot \eta_i} = \frac{U}{q \cdot \eta_i}, \quad \left(kWh \cdot t^{-1}\right)$$

where U is the cell voltage; I is corresponding current; τ is the time of electrolysis; q is the electrochemical equivalent $\left(q = \frac{A}{z \cdot F}\right)$ depending on the atomic weight (A) and valence (z) of the produced metal as well as Faraday constant.

Table: Products of the most massive electrolyses in the world and corresponding energy consumption

Product (Year)	Type of electrolysis*	World's annual production 1000 t	Average specific energy consumption kWh/t	Total energy consumption 1000 kWh
Al (2002)	MSE	29,000	13,500	391,000
Mg (2003)	MSE	340	15,000	5,100
Cu (2003)	ER	11,200	300	3,360
Cu (1999)	EW	1,300	2,000	2,600
Zn (2005)	EW	6,400	3,500	22,400

Ag (2005)	ER	18,300	100	1,830
Pb (2003)	ER	1,320	150	200
Ni (2003)	EW	504	4,000	2,020
Na (2001)	MSE	340	9,000	3060
Li (2003)	MSE	95	–	–
Cd (2003)	EW	16	–	–
Au (2005)	ER	2,5	–	–
Cl_2 (1999)	EW	36,500	3,000	110,000

* - MSE – Molten Salt Electrolysis, ER – Electrorefination, EW – Electrowining

(F); η_i is the current efficiency. According to equation $w = \frac{W}{m} = \frac{U \cdot I \cdot \tau}{I \cdot \tau \cdot q \cdot \eta_i} = \frac{U}{q \cdot \eta_i}, \left(kWh \cdot t^{-1}\right)$, the metal produced by electrolysis can be cheaper if the cell voltage (U) is lower and the electrochemical equivalent and the current efficiency are higher.

Electrochemical equivalent (q) is a characteristic value for each metal and varies only by valence state. Lower energy consumption can be achieved in the case of the lowest valence state of metal ions. For example, electrowining of copper by Cu^+ ions should be as twice efficient as by Cu^{2+} ions.

Current efficiency (η_i) lowers when other electrochemical processes occur or some other energy losses happen. The average current efficiencies of electrometallurgical processes are in the range of 85-95 %. To increase the energy savings, prevention of the above energy losses should be provided. To increase current efficiency, the following factors should be considered:

- High enough concentration of basic metal ions in the electrolyte to avoid deposition of metal impurities on the cathode,
- Pure electrolyte, i.e. minimal concentration of impurity metal ions,
- Optimal acidity of the electrolyte to improve its electrical conductivity,
- Higher working current density at which current efficiency is higher,
- Optimal working temperature of electrolysis. At higher temperatures electrical conductivity of the electrolyte and solubility of metal ions are higher as well as the diffusion of metal ions is faster. On the other hand, the over potential of some undesirable reactions (such as hydrogen evolution) is lower. Thus, the optimal temperature at which current efficiency is the highest has some intermediate value,
- The presence of surface active substances to provide flat surface and to avoid inclusion of electrolyte and anodic slime particles on the metal deposit,
- Avoiding short circuit between electrodes and other losses of energy through the conductors in the cell.

Cell voltage (U) is the main parameter which directly affects the energy consumption of the electrolytic process. The lower the cell voltage is, the lower the specific energy consumption of the electrolysis. But, the cell voltage cannot be lowered infinitely. So, the electrolysis has a minimal value of the cell voltage equivalent to the energy needed for the electrode reactions, i.e. summary chemical reaction. This is known as a theoretical voltage (U_S) of the cell. The cell voltage (U) can be decreased by:

- Proper choice of the electrode materials on which electrode (cathodic and anodic) occurs by the lowest over potential,
- Proper cell design that provides minimal energy losses and
- Knowledge on the cell voltage structure.

The knowledge on the voltage components is very important in order to take action for decreasing them and consequently to decrease overall cell voltage. Beside the energy for the electrode reactions (anodic, η_a and cathodic, η_c over potential), the applied over potential contains other components such as Ohmic drop (IR) as result of electrical resistance of the electrolyte, energy to surmount diffusion limitations (diffusion over potential, η_d) energy for electrocrystallization and voltage as result of the resistance of the metal conductors, contacts, etc. The cell voltage structure can be expressed by the following equation:

$$U = U_s + |\eta_k| + |\eta_a| + \sum |\eta_d| + |\eta_{kr}| + IR$$

Except the theoretical voltage US, which is a thermodynamic non-variable value, all other components can be varied (decreased) by controlling the electrolysis parameters.

In table the voltage structure and the specific energy consumption for electrorefination of copper and electrowining of copper are shown.

Table: Cell voltage U, V structure of the copper electrorefination and copper and zinc electrowining

Cell Voltage Components	**Electrorefining**	**Electrowining**	
	Copper (20÷35 mA·cm^{-2})	Copper (30 mA·cm^{-2})	Zinc (45 mA·cm^{-2})
Theoretical voltage (U_S)	0	0.89	1.99
Overpotential:			
anodic	0.01	0.60	0.60
cathodic	0.08	0.05	0.30
Ohmic drop of the electrolyte ($I \times R$)	0.13	0.13	0.30
Ohmic drop in the metal conductors	0.02	0.05	0.02
Ohmic drop in the contacts	0.06	-	0.05
Total	0.25	1.72	3.26

It is noticeable that the electrorefination process consumes considerably less energy than electrowining. All other components of the cell voltage have similar values in both cases, except for the theoretical voltage. This difference in the cell voltage is due to the different nature of both processes. So, in the case of electrorefination there is not chemical transformation of the matter, because both, anodic and cathodic reactions are the same, occurring in opposite direction. Theoretical cell voltage defined as a sum of anodic and cathodic standard electrode potential, in the case of copper electrorefination is zero:

$$U_S = E_a^o + E_c^o = E_{Cu}^o + \left(-E_{Cu}^o\right) = 0$$

In the case of electrowining, chemical transformation of the matter occurs, for example, ZnSO4 transforms to Zn. Minimal energy needed for this transformation is above the theoretical cell voltage. Theoretical cell voltage for copper electrowining is 0.89 V, while the corresponding value for zinc electrowining is even 1.99 V. Thus, theoretical energy needs are higher in the case of electrowining than those in the case of electrorefining. This implies higher specific energy consumption for electrowining compared to electrorefining process.

Analyzing the column for specific energy consumption in the table, it is noticeable that the electrolysis in molten salts is the largest energy consumer. For illustration, electrolysis parameters for zinc electrowining in aqueous electrolyte and aluminium electrowining in molten salt are given in table.

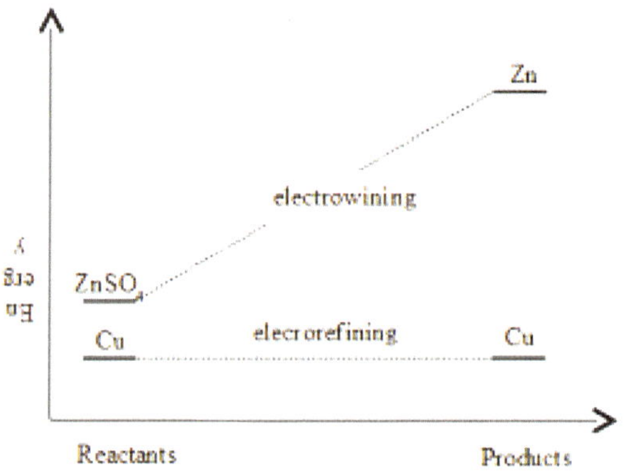

Comparison of the energy change of electrorefination and electrowining

Table: Comparison of the electrolytic parameters of the aqueous electrowining of zinc and molten salt electrowining of aluminium

	U_S/V	U/V	i/mA-cm^{-2}	η_i	T/°C	w/ kWh-t^{-1}
Zinc	1.99	3.3	40 – 70	0.90	35 – 40	3.500
Aluminium	2.89	4.5	6000	0.85	950	15.000

Specific energy consumption is more than five times higher for molten salt electrolysis of aluminium. This is a result of the following factors:

- Higher theoretical cell voltage, due to considerably higher negative value of the standard electrode potential of aluminium and other over potentials;
- High working temperature in the electrochemical cell (~950° C). To maintain this temperature, large quantity of energy is needed i.e. the electrolysis should be performed at higher current densities, even 100 times higher than corresponding ones during aqueous electrowining. These high current densities cause high specific energy consumption;
- Interaction/oxidation of the produced metal with the molten salt which causes decrease of the current efficiency.

Environmental Aspects

Compared to the pyrometallurgical metal production, electrometallurgy is considered as a "clean technology". Theoretically, electrolysis should not pollute the environment due to closed loop of the electrolyte cycling.

Nevertheless, there are several hot spots in electrometallurgical practice. Shown in figure is the scheme of the pollution sites in the electrometallurgical production of metals.

Mineral processing is the main source of the pollution. The minerals treatment from ore to pure electrolyte produces considerable quantity of solid, liquid and gaseous pollutants such as mine overburden, tailing ponds, wastewater, gases, smoke and dust. All of them contain heavy metals and are considered as hazardous waste.

The electrolysis plant is less polluting section compared to the mineral processing. There are two hot spots in the electrolysis plant:

- Evaporation of the electrolyte and gaseous electrode products and
- Wastewaters.

During electrolysis, different gaseous products evolve and evaporation of the electrolyte occurs. Mainly oxygen and/or hydrogen and rarely chlorine or fluorine as gaseous products appears, while fine electrolyte drops evaporate during electrolysis. Proper ventilation has to be installed over the electrolytic cells to take away these pollutants and they must be further treated. In the case of molten salt electrolysis, the gaseous phase is more hazardous due to the presence of fluorine and gaseous fluorine compounds or chlorine and chlorine compounds.

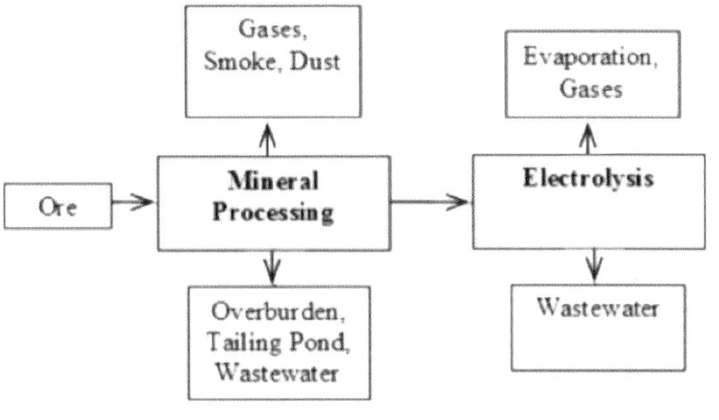

Figure: Scheme of the pollution sites in the electrometallurgical production of metals

Wastewater originates by the change of the used electrolyte and by the panning of the produced metal after electrolysis. This wastewater is acidic, containing heavy metals and different kind of anions. Wastewaters treatment includes neutralization process, performed by chemical deposition, extractive electrolysis and ion exchange in columns.

References

- Extractive-metallurgy-81854: britannica.com, Retrieved 26 April 2018
- What-is-hydrometallurgy: wisegeek.com, Retrieved 23 June 2018
- Hydrometallurgy-2485: corrosionpedia.com, Retrieved 16 July 2018
- Non-ferrous-metal-5260: corrosionpedia.com, Retrieved 15 July 2018
- Physical-metallurgy, metallurgy, inorganic-chemistry: tutorvista.com, Retrieved 25 April 2018

3

Metal Extraction

Metal extraction involves the practice of removing metals from an ore through physical, chemical and electrolytical processes and then refining them into a purer state. The topics elaborated in this chapter on the concentration of ore, isolation of metal from concentrated ore, purification of metal, etc. will help in providing a better perspective about metal extraction.

Concentration of Ore

An ore of a metal is an impure source, this means other than the metal it consists of many other impurities such as sand, grit, clay, rocks etc. These impurities are collectively known as gangue. Gangue is generally commercial valueless and we must separate it from the ore. This entire process is the concentration of ores or even dressing or benefaction.

Hydraulic Washing

Hydraulic Washing is based on the difference in gravity of the particles of the gangue and the ore. So, it is considered as a type of gravity separation. During this process, an upward stream of running water is used to wash the powdered ore. The lighter gangue particles are washed leaving the heavier ore particles. Hydraulic washing is used for ores that have tin or lead, as they are heavier than the gangue.

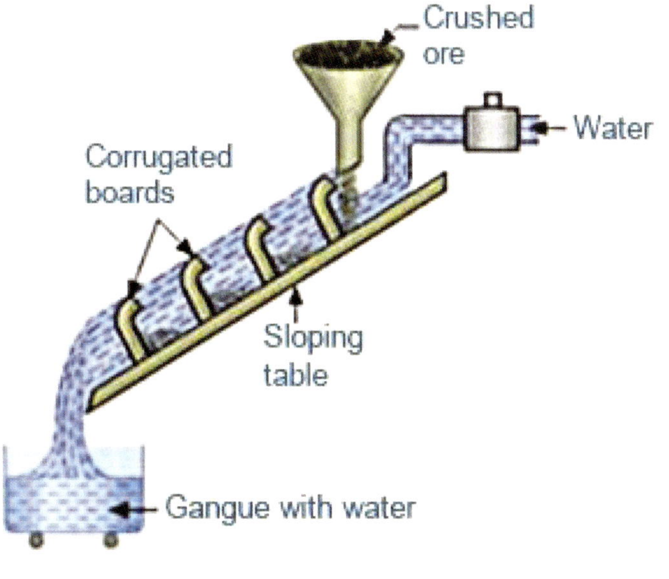

Figure: Hydraulic washing

Magnetic Separation

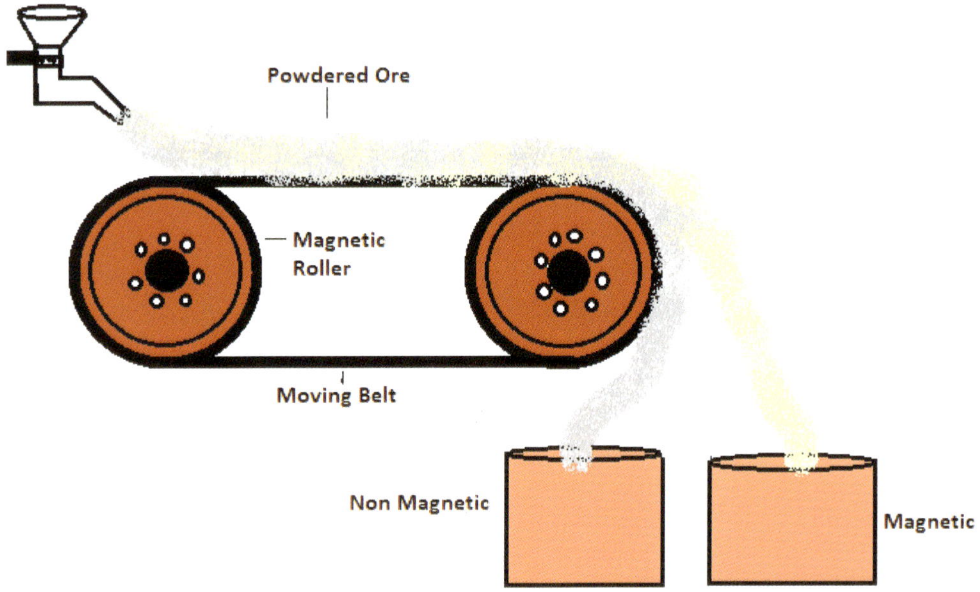

Magnetic separation is a process used to separate materials from those that are less or nonmagnetic. All materials have a response when placed in a magnetic field, although with most, the effect is too slight to be detected. The few materials that are strongly affected (magnetised) by magnetic fields are known as "Ferromagnetics", those lesser (though noticeably) affected are known as "Paramagnetics".

Ferromagnetics require relatively weak magnetic fields to be attracted and devices to separate these materials usually have magnets that are permanently magnetised ("Permanent" magnets do not require electricity to maintain their magnetic fields). Paramagnetics require stronger magnetic fields and these can only be achieved and maintained by "electro" magnets (large wire coils around an iron frame – current is continuously passed through the coils creating the magnetic field within the iron. The field is concentrated across an air gap in the circuit).

Both ferromagnetic (low intensity) and paramagnetic (high intensity) separation devices (Labora tory Magnetic Separator) may be operated with dry solids or with solids in pulp form.

Magnetic separation has two major applications in mineral processing plants:

- The removal of tramp iron (which would deleteriously affect subsequent processes) from an ore stream. This is usually achieved by a low intensity magnet suspended above, or at the head of, a conveyor.
- Separation of magnetic minerals from less magnetic minerals.

This table shows the more common magnetic minerals and the range of field intensities that may be required to separate them.

Some Common Magnetic Minerals

	Mineral	Formula	Field Strength*
Ferromagnetic	Magnetic	Fe_3O_4	→ 1
	Pyrrhotite	Fe_7S_8	0.5 - 4
Paramagnetic	Ilmenite	$FeTiO_3$	8 – 16
	Siderite	$FeCO_3$	9 – 18
	Chromite	$FeCr_2O_4$	10 – 16
	Hematite	Fe_2O_3	12 – 18
	Wolframite	$(Fe,Mn)WO_4$	12 – 18
	Tourmaline		16 – 20

Froth Flotation

Froth flotation is a process for selectively separating hydrophobic materials from hydrophilic. This is used in mineral processing, paper recycling and waste-water treatment industries. Historically this was first used in the mining industry, where it was one of the great enabling technologies of the 20th century. It has been described as "the single most important operation used for the recovery and upgrading of sulfide ores". The development of froth flotation has improved the recovery of valuable minerals, such as copper- and lead-bearing minerals. Along with mechanized mining, it has allowed the economic recovery of valuable metals from much lower grade ore than previously.

Industries

POES

Froth flotation to separate plastics

Froth flotation cells to concentrate copper and nickel sulfide minerals, Falconbridge, Ontario.

Froth flotation is a process for separating minerals from gangue by taking advantage of differences in their hydrophobicity. Hydrophobicity differences between valuable minerals and waste gangue are increased through the use of surfactants and wetting agents. The selective separation of the minerals makes processing complex (that is, mixed) ores economically feasible. The flotation process is used for the separation of a large range of sulfides, carbonates and oxides prior to further refinement. Phosphates and coal are also upgraded (purified) by flotation technology.

Prior to 1907, nearly all the copper mined in the US came from underground vein deposits, averaging 2.5 percent copper. By 1991, the average grade of copper ore mined in the US had fallen to only 0.6 percent.

Waste Water Treatment

The flotation process is also widely used in industrial waste water treatment plants, where it removes fats, oil, grease and suspended solids from waste water. These units are called dissolved air flotation (DAF) units. In particular, dissolved air flotation units are used in removing oil from the wastewater effluents of oil refineries, petrochemical and chemical plants, natural gas processing plants and similar industrial facilities.

Paper Recycling

Froth flotation is one of the processes used to recover recycled paper. In the paper industry this step is called deinking or just flotation. The target is to release and remove the hydrophobic contaminants from the recycled paper. The contaminants are mostly printing ink and stickies. Normally the setup is a two-stage system with 3,4 or 5 flotation cells in series.

Principle of Operation

Before froth flotation can work, the ore to be treated is reduced to fine particles by crushing and grinding (a process known as comminution) so that the various minerals exist as physically separate grains. This process is known as *liberation*. The particle sizes are typically less than 0.1 mm (100 μm), but sometimes sizes smaller than 7–10 μm are required. There is a tendency for the liberation size of the minerals to decrease over time as the ore bodies with coarse mineral grains that can be separated at larger sizes are depleted and replaced by ore bodies that were formerly considered too difficult.

In the mining industry, the plants where flotation is undertaken to concentrate ore are generally known as *concentrators* or *mills*.

For froth flotation, the ground ore is mixed with water to form a slurry and the desired mineral is rendered hydrophobic by the addition of a surfactant or *collector* chemical (although some mineral surfaces are naturally hydrophobic, requiring little or no addition of collector). The particular chemical depends on the nature of the mineral to be recovered and, perhaps, the natures of those that are not wanted. As an example, sodium ethyl xanthate may be added as a collector in the selective flotation of galena (lead sulfide) to separate it from sphalerite (zinc sulfide). This slurry (more properly called the *pulp*) of hydrophobic particles and hydrophilic particles is then introduced to tanks known as *flotation cells* that are aerated to produce bubbles. The hydrophobic particles attach to the air bubbles, which rise to the surface, forming a froth. The froth is removed from the cell, producing a concentrate ("con") of the target mineral.

Frothing agents, known as *frothers*, may be introduced to the pulp to promote the formation of a stable froth on top of the flotation cell.

The minerals that do not float into the froth are referred to as the *flotation tailings* or *flotation tails*. These tailings may also be subjected to further stages of flotation to recover the valuable particles that did not float the first time. This is known as *scavenging*. The final tailings after scavenging are normally pumped for disposal as mine fill or to tailings disposal facilities for long-term storage.

Froth flotation efficiency is determined by a series of probabilities: those of particle–bubble contact, particle–bubble attachment, transport between the pulp and the froth, and froth collection into the product launder. In a conventional mechanically-agitated cell, the void fraction (i.e. volume occupied by air bubbles) is low (5 to 10 percent) and the bubble size is usually greater than 1 mm. This results in a relatively low interfacial area and a low probability of particle–bubble contact. Consequently, several cells in series are required to increase the particle residence time, thus increasing the probability of particle–bubble contact.

Flotation is normally undertaken in several stages to maximize the recovery of the target mineral or minerals and the concentration of those minerals in the concentrate, while minimizing the energy input.

Flotation Stages

Roughing

The first stage is called *roughing*, which produces a *rougher concentrate*. The objective is to remove the maximum amount of the valuable mineral at as coarse a particle size as practical. The finer an ore is ground, the greater the energy that is required, so it makes sense to fine grind only those particles that need fine grinding. Complete liberation is not required for rougher flotation, only sufficient liberation to release enough gangue from the valuable mineral to get a high recovery.

The primary objective of roughing is to recover as much of the valuable minerals as possible, with less emphasis on the quality of the concentrate produced.

In some concentrators, there may be a *preflotation* step that precedes roughing. This is done when there are some undesirable materials, such as organic carbon, that readily float. They are removed first to avoid them floating during roughing (and thus contaminating the rougher concentrate).

Cleaning

The rougher concentrate is normally subjected to further stages of flotation to reject more of the undesirable minerals that also reported to the froth, in a process known as *cleaning*. The product of cleaning is known as the *cleaner concentrate* or the *final concentrate*.

The objective of cleaning is to produce as high a concentrate grade as possible.

The rougher concentrate is often subject to further grinding (usually called *regrinding*) to get more complete liberation of the valuable minerals. Because it is a smaller mass than that of the original ore, less energy is needed than would be necessary if the whole ore were reground. Regrinding is often undertaken in specialized *regrind mills*, such as the IsaMill, designed to further reduce the energy consumed during regrinding to finer sizes.

Scavenging

The rougher flotation step is often followed by a *scavenger* flotation step that is applied to the rougher tailings. The objective is to recover any of the target minerals that were not recovered during the initial roughing stage. This might be achieved by changing the flotation conditions to make them more rigorous than the initial roughing, or there might be some secondary grinding to provide further liberation.

The concentrate from the rougher scavengers could be returned to the rougher feed for refloating or sent to special cleaner cells.

Similarly, the cleaning step may also be followed by a scavenging step performed on the cleaner tailings.

Science of Flotation

To be effective on a given ore slurry, the collectors are chosen based upon their selective wetting of the types of particles to be separated. A good collector will adsorb, physically or chemically, with one of the types of particles. This provides the thermodynamic requirement for the particles to bind to the surface of a bubble. The wetting activity of a surfactant on a particle can be quantified by measuring the contact angles that the liquid/bubble interface makes with it. Another important measure for attachment of bubbles to particles is induction time. The induction time is the time required for the particle and bubble to rupture the thin film separating the particle and bubble. This rupturing is achieved by the surface forces between the particle and bubble.

The mechanisms for the bubble-particle attachment is very complex and consists of three steps, collision, attachment and detachment. The collision is achieved by particles being within the collision tube of a bubble and this is affected by the velocity of the bubble and radius of the bubble. The collision tube corresponds to the region in which a particle will collide with the bubble, with the perimeter of the collision tube corresponding to the grazing trajectory.

The attachment of the particle to the bubble is controlled by the induction time of the particle and bubble. The particle and bubble need to bind and this occurs if the time in which the particle and bubble are in contact with each other is larger than the required induction time. This induction time is affected by the fluid viscosity, particle and bubble size and the forces between the particle and bubbles.

The detachment of a particle and bubble occurs when the force exerted by the surface tension is exceeded by shear forces and gravitational forces. These forces are complex and vary within the cell. High shear will be experienced close to the impeller of a mechanical flotation cell and mostly gravitational force in the collection and cleaning zone of a flotation column.

Significant issues of entrainment of fine particles occurs as these particles experience low collision efficiencies as well as sliming and degradation of the particle surfaces. Coarse particles show a low recovery of the valuable mineral due to the low liberation and high detachment efficiencies.

Selective Adhesion

Froth flotation depends on the selective adhesion of air bubbles to mineral surfaces in a mineral/water slurry. The air bubbles will attach to more hydrophobic particles. The attachment of the bubbles to the surface is determined by the interfacial energies between the solid, liquid, and gas phases. This is determined by the Young-Dupré Equation:

$$\gamma_{lv}\cos\theta = (\gamma_{sv} - \gamma_{sl})$$

where:

- γ_{lv} is the surface energy of the liquid/vapor interface,
- γ_{sv} is the surface energy of the solid/vapor interface,
- γ_{sl} is the surface energy of the solid/liquid interface,
- θ is the contact angle, the angle formed at the junction between vapor, solid, and liquid phases.

Minerals targeted for separation may be chemically surface-modified with collectors so that they are more hydrophobic. Collectors are a type of surfactant that increase the natural hydrophobicity of the surface, increasing the separability of the hydrophobic and hydrophilic particles. Collectors either chemically bond via chemisorption to the mineral or adsorb onto the surface via physisorption.

IMFs and Surface Forces in Bubble-particle Interactions

Collision

The collision rates for fine particles (50 - 80 μm) can be accurately modeled, but there is no current theory that accurately models bubble-particle collision for particles as large as 300 μm, which are commonly used in flotation processes.

For fine particles, Stokes law underestimates collision probability while the potential equation based on Surface Charge overestimates collision probability so an intermediate equation is used.

It is important to know the collision rates in the system since this step precedes the adsorption where a three phase system is formed.

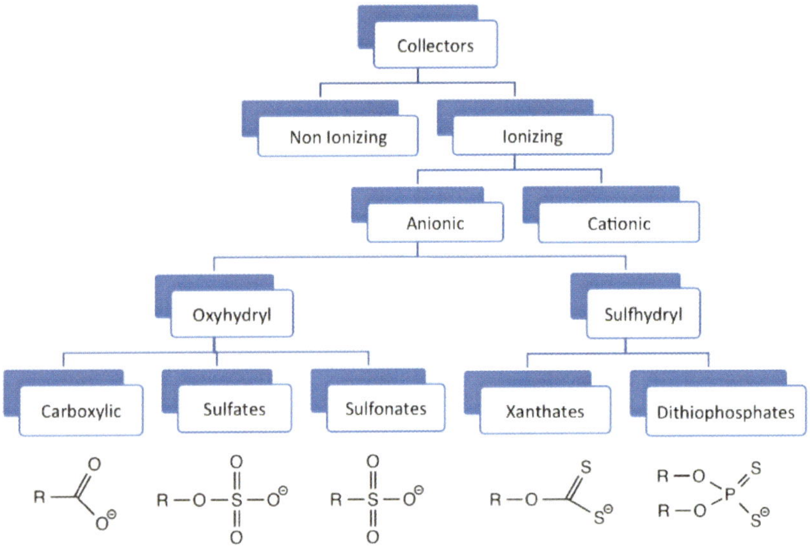

Different types of collectors, or surfactants, used in froth flotation.

Adsorption

The effectiveness of a medium to adsorb to a particle is influenced by the relationship between the surfaces of both materials. There are multiple factors that affect the efficiency of adsorption in chemical, thermodynamic, and physical domains. These factors can range from surface energy and polarity to the shape, size, and roughness of the particle. In froth flotation, adsorption is a strong consequence of surface energy, since the small particles have a high surface area to size ratio, resulting in higher energy surfaces to form attractions with adsorbates. The air bubbles must selectively adhere to the desired minerals to elevate them to the surface of the slurry while wetting the other minerals and leaving them in the aqueous slurry medium.

Particles that can be easily wetted by water are called hydrophilic, while particles that are not easily wetted by water are called hydrophobic. Hydrophobic particles have a tendency to form a separate phase in aqueous media. In froth flotation the effectiveness of an air bubble to adhere to a particle is based on how hydrophobic the particle is. Hydrophobic particles have an affinity to air bubbles, leading to adsorption. The bubble-particle combinations are elevated to the froth zone driven by buoyancy forces.

The attachment of the bubbles to the particles is determined by the interfacial energies of between the solid, liquid, and vapor phases, as modeled by the Young/Dupre Equation. The interfacial energies can be based on the natural structure of the materials, or the addition of chemical treatments can improve energy compatibility.

Collectors are the main additives used to improve particle surfaces. They function as surfactants to selectively isolate and aid adsorption between the particles of interest and bubbles rising through the slurry. Common collectors used in flotation are anionic sulfur ligands, which have a bifunctional structure with an ionic portion which shares attraction with metals, and a hydrophobic portion such as a long hydrocarbon tail. These collectors coat a particle's surface with a monolayer of non-polar substance to aid separation from the aqueous phase by decreasing the adsorbed particle solubility in water. The adsorbed ligands can form micelles around the particles and form small-particle colloids improving stability and phase separation further.

Desorption

The adsorption of particles to bubbles is essential to separating the minerals from the slurry, but the minerals must be purified from the additives used in separation, such as the collectors, frothers, and modifiers. The product of the cleaning, or desorption process, is known as the cleaner concentrate. The detachment of a particle and bubble requires adsorption bond cleavage driven by shear forces. Depending on the flotation cell type, shear forces are applied by a variety of mechanical systems. Among the most common are impellers and mixers. Some systems combine the functionalities of these components by placing them at key locations where they can take part in multiple froth flotation mechanisms. Cleaning cells also take advantage of gravitational forces to improve separation efficiency.

Performance Calculations

Relevant Equations

A common quantity used to describe the collection efficiency of a froth flotation process is *flotation recovery* (*R*). This quantity incorporates the probabilities of collision and attachment of particles to gas flotation bubbles.

$$R = \frac{N_c}{\left(\frac{\pi}{4}\right)(d_p + d_b)^2 Hc}$$

where:

- $N_c = PN_c^i$, which is the product of the probability of the particle being collected (P) and the number of possible particle collisions $\left(N_c^i\right)$.
- d_p is particle diameter.
- d_b is bubble diameter.
- H is a specified height within the flotation which the recovery was calculated.
- c is the particle concentration.

The following, are several additional mathematical methods often used to evaluate the effectiveness of froth flotation processes. These equations are more simple than the calculation for *flotation recovery*, as they are based solely on the amounts of inputs and outputs of the processes.

For the following equations:

- F is the weight percent of feed.
- C is the weight percent concentrate.
- T is the weight percent of tailings.

- c, t, and f are the metallurgical assays of the concentrate, tailings, and feed, respectively

Ratio of feed weight to concentrate weight $\frac{F}{C}$ (unitless)

$$\frac{F}{C} = \frac{c-t}{f-t}$$

Percent of metal recovered (X_R) in wt%

$$X_R = 100\left(\frac{c}{f}\right)\left(\frac{f-t}{c-t}\right)$$

Percent of metal lost (X_L) in wt%

$$X_L = 100 - X_R$$

Percent of weight recovered (X_W) in wt%

$$X_W = 100\left(\frac{C}{F}\right) = 100\frac{f-t}{c-t}$$

Grade-recovery Curves

Grade-recovery curves are useful tools in weighing the trade-off of producing a high grade of concentrate while maintaining as low of a recovery rate as possible, two important aspects of froth flotation. These curves are developed empirically based on the individual froth flotation process of a particular plant. As the curves are shifted in the positive x-direction (to the right) and the positive y-direction (upward) the performance of the froth flotation process is regarded as improving. A disadvantage to these curves is that they can only compare the grade-recovery relations of a specific feed grade and feed rate. If a company has a variance of feed grades and rates used (an extremely common occurrence) in their froth flotation process, grade-recovery curves for every pairing of feed grade and recovery rate would have to be constructed in order to provide meaningful information to the plant.

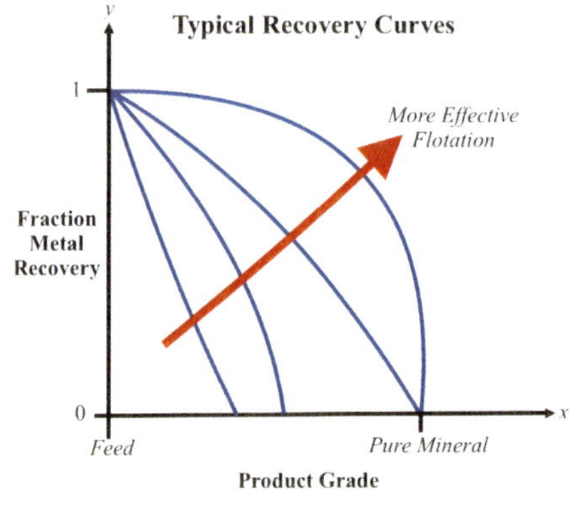

Example grade-recovery relationships seen in froth flotation. Shifts in the curves represent changes in flotation effectiveness.

Flotation Equipment

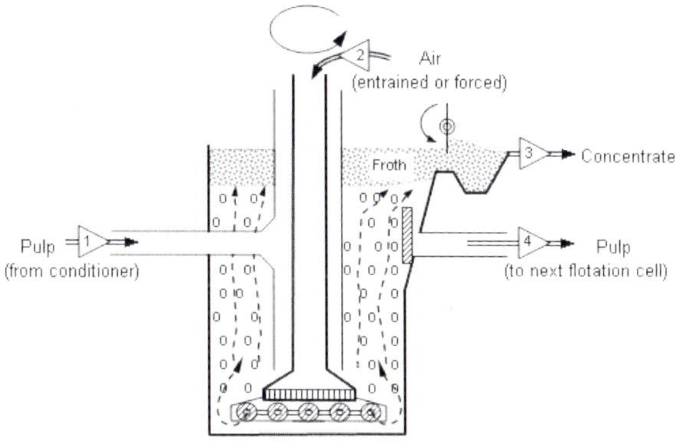

Froth flotation cell.

In diagram numbered triangles show direction of stream flow. A mixture of ore and water called pulp enters the cell from a conditioner, and flows to the bottom of the cell. Air or nitrogen is passed down a vertical impeller where shearing forces break the air stream into small bubbles. The mineral concentrate froth is collected from the top of the cell, while the pulp flows to another cell.

Flotation can be performed in rectangular or cylindrical mechanically agitated cells or tanks, flotation columns, Jameson Cells or deinking flotation machines. Classified by the method of air absorption manner, it is fair to state that two distinct groups of flotation equipments have arisen:pneumatic and mechanical machines. Generally pneumatic machines give a low-grade concentrate and little operating troubles.

Mechanical cells use a large mixer and diffuser mechanism at the bottom of the mixing tank to introduce air and provide mixing action. Flotation columns use air spargers to introduce air at the bottom of a tall column while introducing slurry above. The countercurrent motion of the slurry flowing down and the air flowing up provides mixing action. Mechanical cells generally have a higher throughput rate, but produce material that is of lower quality, while flotation columns generally have a low throughput rate but produce higher quality material.

The Jameson cell uses neither impellers nor spargers, instead combining the slurry with air in a downcomer where high shear creates the turbulent conditions required for bubble particle contacting.

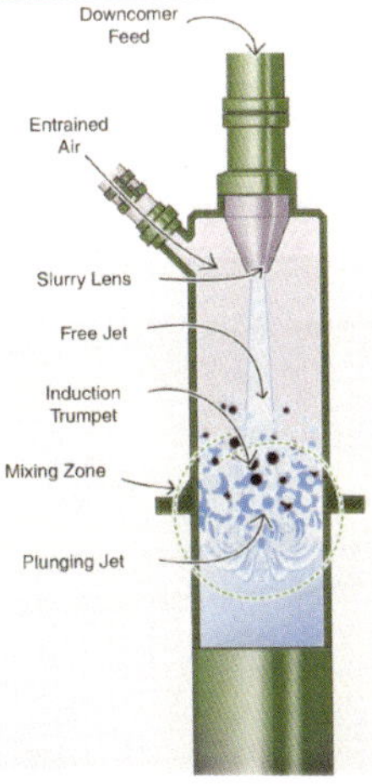

Comparison of the sizes of flotation columns and Jameson Cells with similar capacities.

Mechanics of Flotation

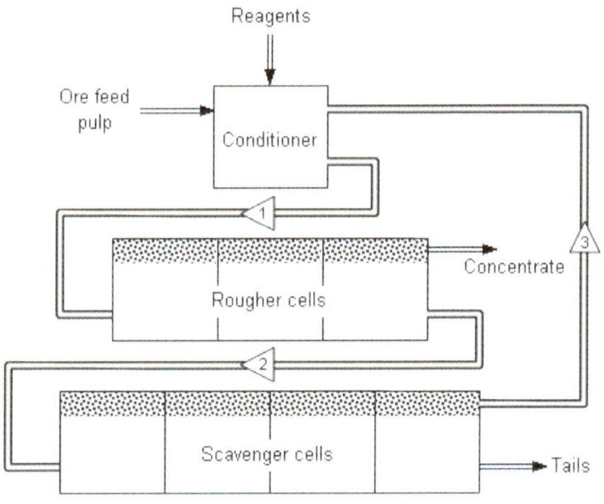

The following steps are followed, following grinding to liberate the mineral particles:

1. Reagent conditioning to achieve hydrophobic surface charges on the desired particles.
2. Collection and upward transport by bubbles in an intimate contact with air or nitrogen.
3. Formation of a stable froth on the surface of the flotation cell.
4. Separation of the mineral laden froth from the bath (flotation cell).

Simple Flotation Circuit for Mineral Concentration

Numbered triangles show direction of stream flow. Various flotation reagents are added to a mixture of ore and water (called pulp) in a conditioning tank. The flow rate and tank size are designed to give the minerals enough time to be activated. The conditioner pulp is fed to a bank of rougher cells which remove most of the desired minerals as a concentrate. The rougher pulp passes to a bank of scavenger cells where additional reagents may be added. The scavenger cell froth is usually returned to the rougher cells for additional treatment, but in some cases may be sent to special cleaner cells. The scavenger pulp is usually barren enough to be discarded as tails. More complex flotation circuits have several sets of cleaner and re-cleaner cells, and intermediate re-grinding of pulp or concentrate.

Chemicals of Flotation

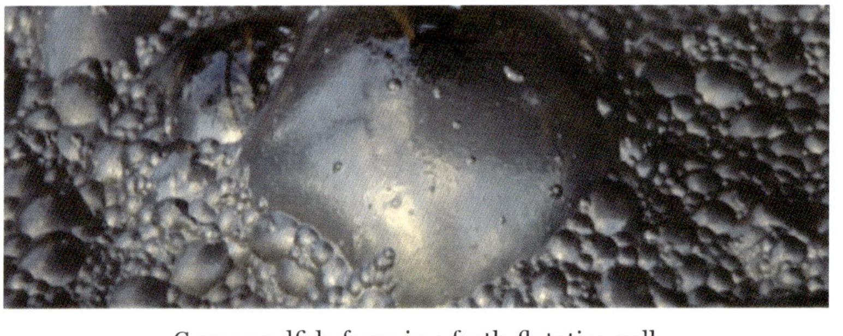

Copper sulfide foam in a froth-flotation cell

Collectors

For many ores (e.g. those of Cu, Mo, W, Ni), the collectors are anionic sulfur ligands. Particularly popular are xanthate salts, including potassium amyl xanthate (PAX), potassium isobutyl xanthate (PIBX), potassium ethyl xanthate (KEX), sodium isobutyl xanthate (SIBX), sodium isopropyl xanthate (SIPX), sodium ethyl xanthate (SEX). Other collectors include related sulfur-based ligands: dithiophosphates, dithiocarbamates. Still other classes of collectors include the thiourea thiocarbanilide. Fatty acids have also been used.

For some minerals (e.g., sylvanite for KCl), fatty amines are used as collectors.

Frothers

A variety of compounds are added to stabilize the foams. These additives include pine oil, various alcohols (methyl isobutyl carbinol (MIBC)), polyglycols, xylenol (cresylic acid).

Modifiers

A variety of other compounds are added to optimize the separation process, these additives are called modifiers. Such species do not interact directly with the mineral, but modify the physical properties of the solution.

- pH modifiers include lime (CaO), Soda ash (Na_2CO_3), Caustic soda (NaOH), sulfuric and hydrochloric acid (H_2SO_4, HCl).
- Anionic modifiers include phosphates, silicates, and carbonates.
- Organic modifiers include the thickeners dextrin, starch, glue, and CMC.

Chemical Method or Leaching

Leaching is a widely used extractive metallurgy technique which converts metals into soluble salts in aqueous media. Compared to pyrometallurgical operations, leaching is easier to perform and much less harmful, because no gaseous pollution occurs. The only drawback of leaching is its lower efficiency caused by the low temperatures of the operation, which dramatically affect chemical reaction rates.

There are a variety of leaching processes, usually classified by the types of reagents used in the operation. The reagents required depend on the ores or pretreated material to be processed. A typical feed for leaching is either oxide or sulfide.

For material in oxide form, a simple acid leaching reaction can be illustrated by the zinc oxide leaching reaction:

$$ZnO + H_2SO_4 \rightarrow ZnSO_4 + H_2O$$

In this reaction solid ZnO dissolves, forming soluble zinc sulfate.

In many cases other reagents are used to leach oxides. For example, in the metallurgy of aluminium, aluminium oxide is subject to leaching by alkali solutions:

$$Al_2O_3 + 3H_2O + 2NaOH \rightarrow 2NaAl(OH)_4$$

Leaching of sulfides is a more complex process due to the refractory nature of sulfide ores. It often involves the use of pressurized vessels, called autoclaves. A good example of the autoclave leach process can be found in the metallurgy of zinc. It is best described by the following chemical reaction:

$$2ZnS + O_2 + 2H_2SO_4 \rightarrow 2ZnSO_4 + 2H_2O + 2S$$

This reaction proceeds at temperatures above the boiling point of water, thus creating a vapor pressure inside the vessel. Oxygen is injected under pressure, making the total pressure in the autoclave more than 0.6 MPa.

The leaching of precious metals such as gold can be carried out with cyanide or ozone under mild conditions

Vat Leaching

Vat Leaching also known as "agitated tank" leaching, the lixiviant comes into contact with the metal material in large vats or tanks, which can be stirred, enhancing the reaction kinetics. The metal containing solid, such as concentrate, ore, residue, or slag, often undergoes size reduction via crushing and grinding prior to leaching.

One example of a leaching process that uses vat leaching is gold cyanidation, the process of extracting gold from low-grade ores. In this process, a dilute solution of sodium cyanide (NaCN) is used to leach the Au into solution. The concentration is usually 0.01-0.05 % cyanide, or 100-500ppm. It should also be noted that the alkalinity of this solution must be high enough such that hydrogen cyanide is not produced, which is a very toxic gas. This is more important when sulfide minerals are in the ore, because they will oxidize, consuming oxygen and generating acid. The pH can be kept above 10, by adding a base such as lime (CaO) to the solution, at a minimum concentration of 0.04%, when NaCN is 0.01%. Other additives include lead nitrate solution (at a ratio of 2:3 to the NaCN solution) in order to minimize cyanide consumption in the reaction. Based on different variations of gold cyanide leaching and the starting material, different concentrations and ratios of the solvents and additives are used.

Heap Leaching

Ore is crushed, agglomerated and charged to a heap, in a container lined with an impervious layer. Lixiviant is then irrigated on top of the heap, diffused through the heap, and is eventually collected and pumped out for further processing.

Usually, this method is used on low grade ores that cannot be economically processed via other means. Heap leaching is successfully used to extract gold, silver, copper, nickel, and uranium. The largest heap leach for gold is at the Yanacocha mine in Peru, second only to Veladero in Argentina. Heap leaching is used in tandem with electrowinning, for copper extraction; approximately 16% of the world's total copper production is extracted in this way. The Radomiro Tomic mine in Chile is a prime example of the success of copper heap leaching, followed by solvent extraction and electrowinning.

Heap leaching technology is important to the future of metal recovery, since it can be applied to extract precious and base metals sitting in abandoned tailings, and/or waste management sites. With emerging technologies and development requiring more of our natural resources, near 100% recovery becomes extremely important.

In-situ Leaching

Holes are drilled into the deposit, in order to make pathways for the solution to penetrate. The lixiviant is then pumped into the deposit, making contact with the ore, and the resulting solution is then collected and processed.

Autoclave Leaching

Higher temperature and pressure are used to enhance the rate of the reaction. This aggressive method of leaching is typically used for ores and concentrates that are not amenable to atmospheric leaching, such as sulphides.

Each variation of leaching plays an important role in mining and metal recovery, and are used in a variety of applications across the globe.

Isolation of Metal from Concentrated Ore

Isolation of metal from the concentrated ore involves the refining of metals from the compounds it is found with to its original state. The concentrated ore is first converted to an oxide after which metal extraction is done by reducing it from an oxide to metal using a suitable reducing agent. Hence the ores are converted to oxides, so it is ideal for reduction. After doing so the ores will undergo reduction, to give us metal. Therefore the two main steps of extraction of metals are;

- Convert ores to form oxides.
- Reduction of metal oxides.

Convert Ores to Oxides

It is easier to reduce oxides than say sulphides. Oxygen is more electronegative than other elements. So it is easier for oxides to accept an electron pair and get reduced. There are various ways to carry out this step. Let us take a look at them

1. Calcination

This is the process of heating a substance in a limited supply of air or oxygen. We subject the ore to thermal energy (heat) so we can change their chemical state and turn them into oxides. Although we must ensure that we keep the temperatures below the melting point of the metal. Hydroxides and carbonates are usually converted to oxides via this process. Here is how an iron ore is converted to iron oxide, to then later enable reduction

$$Fe_2O_3.xH_2O \rightarrow Fe_2O_3\text{ (solid)} + H_2O\text{ (gas)}$$

2. Roasting

Roasting is another way to convert compounds to oxides. It is a metallurgical process involving gas-solid reactions at elevated temperatures with the goal of purifying the metal component. In roasting the ore is heated up in the furnace in a regular supply of air. Again we ensure that the temperatures are within the melting point of the metal.

Here the oxygen in the air reacts with the sulphide ores to give us oxides. The sulphur in the ore forms sulphur dioxide and separates out. Here is a Zinc Sulphide chemical reaction

$$2ZnS + 3O_2 \rightarrow 2ZnO + 2SO_2$$

Reduction of Metal Oxides

This is the second step in the extraction of metals from their respective ores. Now once we obtain metal oxides from the ores, they must now reduce them. Again in this reduction process, we heat the metal oxide. But this time we add a reducing agent. This reducing ageing is usually Carbon or a carbon compound. This is because they react best with oxygen and are readily and cheaply available.

Now when the carbon reacts with the metal oxide reduction takes place, This is when electron gain or electronation occurs. This means the carbon combines with the oxygen, leaving us with a pure metal. The general reaction for reduction is as follow:

$$M_xO_y + yC \rightarrow xM + CO$$

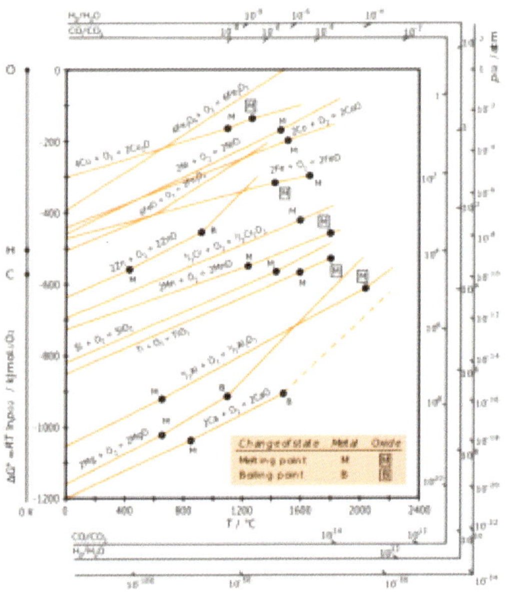

To choose the ideal reducing agent for a certain ore, and to also find out the correct amount of thermal energy requirement we must depend on Giggs Energy interpretations. We represent these graphically using an Ellingham Diagram.

Purification of the Metal

Refining of metals is the removal of impurities from crude metals. After extraction from raw materials, crude metals are between 96 and 99 percent principal metal, the rest being impurities. Crude metals cannot be used by industry at this stage because of inferior physical, chemical, and mechanical properties. The impurities found in crude metals may have value in themselves; the gold and silver recoverable from copper, for example, pay for the entire cost of the refining process.

The three basic refining methods are pyrometallurgical, electrolytic, and chemical. All these methods are based on distinctive properties of the individual elements, such as melting temperature, density, and electronegativity. Pure metals are frequently obtained by employing several refining methods in succession.

Pyrometallurgical Refining

Pyrometallurgical refining, which is carried out at a high temperature in a melt, has a number of variations. Oxidizing refining is based on the tendency of some impurities to form compounds with O, S, Cl, and F that are more stable than compounds of the principal metal and these elements. This technique is used to refine Cu, Pb, Zn, and Sn. For example, when a stream of air is forced through molten copper, oxides are formed by admixtures of Fe, Ni, Zn, Pb, Sb, As, and Sn, since the admixtures have a greater tendency to react with oxygen than copper; the oxides rise to the surface of the tank and are removed.

Fractional recrystallization utilizes the difference in the solubilities of a metallic admixture in solid and liquid phases and the slow diffusion of impurities in the solid phase. This method is used in the production of semiconductor materials and in the preparation of high-purity metals; it is employed in zone melting, plasma metallurgy, the removal of single crystals from a melt, and directed crystallization.

Rectification, or distillation, is based on the difference between the boiling point of the principal metal and that of the impurities. Refining is carried out as a continuous refluxing process in which volatilization and condensation of the fraction being separated are repeated many times. Rectification may be accelerated considerably if it is performed in a vacuum. This method has applications in the removal of Cd from Zn or Zn from Pb, in the separation of Al and Mg, and in Ti metallurgy. Vacuum filtration of a liquid metal through ceramic filters removes suspended solid impurities; the process is used in Sn metallurgy.

When steel is refined in a casting ladle with liquid synthetic slag, the contact surface of the metal and the slag is appreciably greater (because of the mixing) than when the refining processes are carried out in a melting aggregate. This greatly improves the efficiency of desulfurization, dephosphorization, and deoxidation of metals in removing nonmetallic impurities. Inert gases are blown through molten steel in a refining procedure that removes suspended slag particles or solic oxides from the metal. The impurities adhere to gas bubbles and float to the surface of the melt.

Chemical Refining

Chemical refining is based on the different solubilities of a metal and its impurities in acid or alkaline solutions. Impurities gradually accumulate in a solution and are removed by chemical means, such as hydrolysis, cementation, the formation of sparingly soluble compounds, and purification by extraction or ion exchange. The affinage of noble metals is an example of chemical refining. Au is refined in boiling sulfuric or nitric acid; Cu, Ag, and other metallic impurities are dissolved, whereas the purified gold remains as an insoluble residue.

Distillation

Distillation is the process used to separate the substances composing a mixture. It involves a change of state, as of liquid to gas, and subsequent condensation. The process was probably first used in the production of intoxicating beverages. Today, refined methods of distillation are used in many industries, including the alcohol and petroleum industries.

Distillation by Means of Chemical Transport Reactions

Distillation of substances by means of chemical transport reactions as a method for their separation and refining is carried out on the basis of transformation of the main metal into a chemical compound characterised by higher volatility than that of admixtures or other components. In the subsequent dissociation of a compound with low volatility, it is possible to produce distilled metal which is usually characterised by a considerably higher purity in comparison with direct distillation, if this can be carried out.

The principle of distillation by means of chemical transport reactions is based on the reversible reaction between the substance (A) and the gas or vapour (B) with the formation of a gaseous product (C):

$$m\,\mathrm{A}_{[S,L]} + \mathrm{B}_{[G]} \rightleftarrows r\,\mathrm{C}_{[G]}$$

If in one area of equipment this reaction takes place from left to right and in the other area in the opposite direction, in the former part the amount of the solid or liquid phase decreases and in the upper part of equipment increases. This process results in the distillation or transfer of the given substance from one part of equipment to another by means of a chemical transport reaction.

Distillation of Metals by Means of a Reversible Endothermic Exchange Reaction

The metals and elements characterised by the very low vapour pressure at temperatures used in practice, may be distilled by means of a reversible exchange reaction of the following type:

$$m\,\mathrm{Me}_{[\mathrm{cond}]} + n\,\mathrm{MaX}_{m[G]} \underset{T_2}{\overset{T_1}{\rightleftarrows}} m\,\mathrm{MeX}_{n[G]} + n\,\mathrm{Ma}_{[G]}$$

where Me is the non-volatile metal with valency n; Ma is the volatile metal with valency m; X is a halogenide.

The distillation of metals by means of the so-called indirect distillation makes it possible to reduce greatly amount of heat required for the transfer of metals from the solid to gas state. The principle of this method is therefore based on the combination of the strongly endothermic process of direct evaporation and the exothermic reaction. Schematically, these reactions may be described by the following equations:

$$m\,\mathrm{Me}_{[\mathrm{cond}]} \rightleftarrows m\,\mathrm{Me}_{[G]} - m\Delta H_{\mathrm{ev}}$$

$$m\,\mathrm{Me}_{[G]} + n\,\mathrm{Max}_{m[G]} \rightleftarrows m\,\mathrm{MeX}_{n[G]} + n\,\mathrm{Ma}_{[G]} + \Delta H_b$$

where ΔH_{ev} is the change of the enthalpy of evaporation of the metal; ΔH_b in the change of the enthalpy of the exothermic reaction.

The joint occurrence of these reactions results in the transfer of metal Me into a gas compound in the form of MeX;

$$m\,\mathrm{Me}_{[\mathrm{cond}]} + n\,\mathrm{Max}_{m[G]} \underset{T_2}{\overset{T_1}{\rightleftarrows}} m\,\mathrm{MeX}_{n[G]} + n\,\mathrm{Ma}_{[G]} - (m\Delta H_{\mathrm{vb}} - \Delta H_b)$$

Therefore, this process is less endothermic than direct evaporation and, consequently, it takes place at a considerably lower temperature. This process is a reversible and, consequently, after transport in cooling to temperature T_2 the re-condensed non-volatile metal (Me) and the vapours of the volatile halogenide of the metal (MaX) form again in a new area as the product of the reaction. In order to ensure that the reaction is reversible, both the metals and halogenides must not form mutually stable compounds in the conditions of the reaction taking place.

The distillation of metals by their stable halogenides is realised in practice by passing the vapours of the halogenides of the volatile metal above the non-volatile metal. The resultant transport gas mixture cools down in a different area of the apparatus and this results in a reverse process and the non-volatile metal condenses completely or partially free from halogenides or contains a certain amount of these halogenides (depending on the properties of the reacting compounds and the process conditions).

Since the reaction of indirect distillation is endothermic, increase of temperature supports the occurrence of this reaction, and a decrease of pressure increases the reaction rate. The efficiency of separation and refining of metals in indirect distillation depends on the affinity of the metal and individual admixtures for the individual halogenides, i.e. on the decrease of the Gibbs energy with the formation of halogenides and on the concentration of each element in the alloy. A suitable example is the refining of beryllium, titanium and nickel:

$$\mathrm{Be}_{[S]} + 2\,\mathrm{NaCl}_{[G]} \rightleftarrows \mathrm{BeCl}_{2[G]} + 2\,\mathrm{Na}_{[G]}$$

$$\mathrm{Ti}_{[S]} + 4\,\mathrm{NaCl}_{[G]} \rightleftarrows \mathrm{TiCl}_{4[G]} + 4\,\mathrm{Na}_{[G]}$$

$$\mathrm{Ni}_{[G]} + 2\,\mathrm{HCl}_{[G]} \rightleftarrows \mathrm{NiCl}_{2[G]} + \mathrm{H}_{2[G]}$$

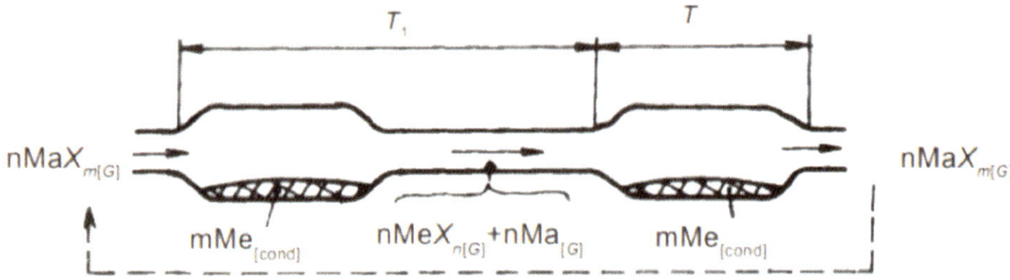

Figure: Distilation of metal [Me] by a reversible endothermic ($T_1 > T_2$):

$$m\,\mathrm{Me}_{[cond]} + n\,\mathrm{Ma}\,X_{m[G]} \rightleftarrows m\,\mathrm{Me}\,X_{n[G]} + n\,\mathrm{Ma}_{[G]}$$

In the vacuum conditions, the activity of these reactions is quite high in the range T_1 at the evaporation temperature and the temperature of condensation of chlorides at 800 °C and in the range T_2 at the temperature of the reaction of 1000 to 1250 °C. It should be stressed that at the reaction temperature T_2 = 1400 °C, the amount of distilled titanium is 500 times larger than the amount which would have been obtained in simple evaporation of metallic titanium in the same temperature conditions.

Distillation of Metals through Sub-compounds

Another method of obtaining a high degree of purification and separation of metals by means of chemical transport reactions is distillation or sublimation of sub-compounds of metals. The sub-compounds are highly volatile non-saturated compounds of metals with a reduced valency. The compounds with reduced valency form in cases in which not all valency electrons of the atom take part in the chemical reaction, and only those characterised by the weakest bonding with the core participate, i.e. those characterised by the lowest value of the ionisation potential. The sub-compounds form at higher temperatures, mainly in the inert atmosphere or in vacuum, with the simultaneous reaction of the atom of the metal with some of the chemical compounds of the same metal, for example, halogenide:

$$\mathrm{Me} + \mathrm{MeX} \underset{T_2}{\overset{T_1}{\rightleftarrows}} \mathrm{Me_2X}$$

In this case, the process becomes less endothermic in comparison with simple sublimation or distillation. The reaction may be described as the sum of two reactions:

$$\mathrm{Me}_{[\mathrm{cond}]} \rightleftarrows \mathrm{Me}_{[G]} - \Delta H_{ev}$$

$$\mathrm{Me}_{[G]} + \mathrm{MeX}_{[G]} \rightleftarrows \mathrm{Me_2X}_{[G]} + \Delta H_b$$

$$\mathrm{Me}_{[\mathrm{cond}]} + \mathrm{MeX}_{[G]} \underset{T_2}{\overset{T_1}{\rightleftarrows}} \mathrm{Me_2X}_{[G]} - \left(\Delta H_{ev} - \Delta H_b\right)$$

On the whole, the reaction remains endothermic but in comparison with direct distillation it is less endothermic so that the distillation process may be realised at a considerably lower temperature. The previously mentioned reaction is reversible. At a high temperature (T_1) reaction takes place in the direction of formation of volatile sub-compounds. After cooling to temperature (T_2), the sub-compound formed in this manner dissociates into a metal and its compound with a high valency.

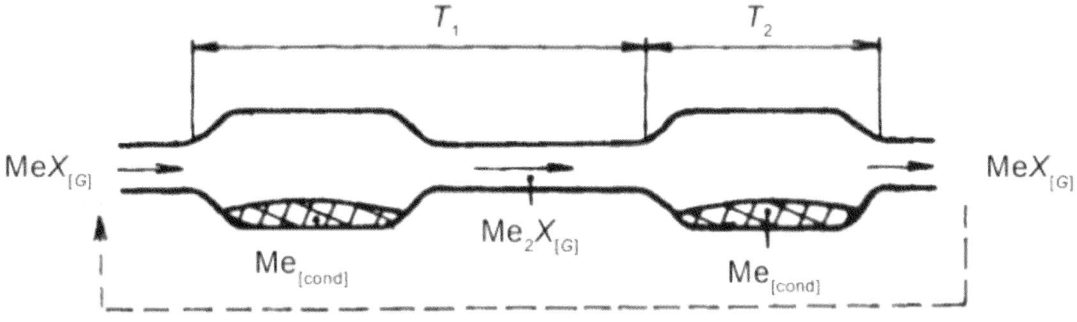

Figure: Diagram of distillation of metal through sub-compounds, transfer of metal to the cold zone,

$T_1 > T_2$: $Me_{[cond]} + Me\,X_{[G]} \rightleftarrows Me_2\,X_{[G]}$

In this case, the carrier of distilled metal is the volatile compound with reduced valency (sub-compound) of the same metal. An increase of temperature and a decrease of pressure accelerate the reaction of formation of the sub-compound and, consequently, the process of transport distillation of the metal. The formation of compounds of the univalent subchloride AlCl at higher temperatures ($T_1 \approx 1000$ °C) is a consequence of the chemical reaction of metallic aluminium with $AlCl_3$ compound with a higher valency:

$$2\,Al_{[cond]} + AlCl_{3[G]} \underset{T_2}{\overset{T_1}{\rightleftarrows}} 3\,AlCl_{[G]}$$

The compounds of the univalent metal usually exist in the gas state (in the inert atmosphere or in vacuum). With a decrease of temperature ($T_2 \approx 600$ °C), the equilibrium of the reaction of the formation of sub-compounds of metals is displaced the left and these compounds dissociate into the transported metal (aluminium) and its compounds with a higher valency ($AlCl_3$).

The capacity of aluminium to form volatile compounds with a lower valency, in particular, subfluorides and subchlorides, may be utilised for the separation of aluminium from other metals and for deep refining to remove admixtures. The refining of aluminium by means of subcompounds takes place in special distillation columns in accordance with the reactions:

$$AlF_{3[G]} + 2Al_{[L]} \xrightarrow{T_i} 3\,AlF_{[G]} \xrightarrow{T_2} 2\,Al_{[L]} + AlF_{3[G]}$$

$$AlCl_{3[G]} + 2\,Al_{[L]} \xrightarrow{1000\,^0C} 3\,AlCl_{[G]} \xrightarrow{600\,^\circ C} 2\,Al_{[S]} + AlCl_{3[G]}$$

Other examples:

$$Si_{[S]} + SiCl_{4[G]} \rightleftarrows 2\,Si\,Cl_{2[G]}$$

$$Ti_{[S]} + TiCl_{4[G]} \rightleftarrows 2\,Ti\,Cl_{2[G]}$$

The conditions for obtaining a pure metal by distillation of these metals through subhalogenides are:

- The application of the initial metal with the minimum content of admixtures;
- Efficient purification, for example, by double or triple sublimation of initial halogenides;
- The application of high quality corrosion-resisting materials for the manufacture of components of equipment with which the refined metal and its halogenides come into contact.

Table shows examples of the transport reactions by means of chlorides and subchlorides of metals for different types of the chemism of the reactions:

1. Transport of metals by means of sub-compounds
2. Transport of chlorides and oxychlorides of metals
3. Transport of oxides and other compounds.

Distillation by Synthesis and Dissociation of Volatile Compounds: Iodide Reaction

In 1925, van Arkel proposed the so-called iodide process for refining of refractory metals.

The principle of the method is based on the following: the volatile iodide of a metal is heated and evaporated in a vessel 6 produced from glass with a higher melting point in which the thin W-wire (fibre) 4 is suspended and secured at points 5 and 8 between tungsten bars 1, 2, brazed into the vessel. The passage of electric current through W-wire results in heating of the wire to a high temperature, $t_2 > 1400$ °C, accompanied by the dissociation of the iodide. Consequently, the metal settles on the surface of the wire in pure form. The iodide of the metal is obtained at a different temperature in a different area of the same vessel.

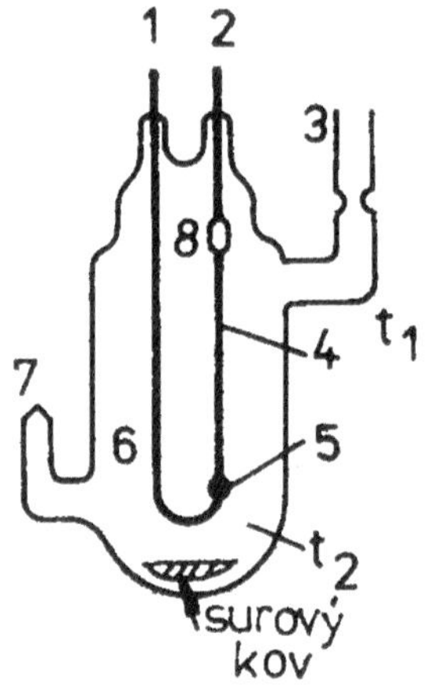

Figure: Diagram of the reaction vessel for the iodide reaction of metals

The initial powder metal and the iodide are introduced into the vessel through the pipe 7 which is subsequently sealed by melting. Vacuum is produced in the vessel and the pipe 3 is also sealed by melting. The entire system is heated in an electric furnace to temperature $t_1 > 200$ °C at which the initial metal and iodide reacted to form the metal tetraiodide and evaporate. Temperature t_1 is considerably lower than the temperature to which the tungsten wire is heated and at which thermal dissociation of the iodide takes place. The transport dissociation iodide reaction may be expressed by the following schematic equation:

$$Me_{[S]}+21_{2[G]}\underset{T_2}{\overset{T_1}{\rightleftarrows}}Mel_{4[G]} \text{ and } T1 < T2.$$

In this case, the metal is transported from the colder to warmer zone. The reaction of formation of a volatile intermediate product – carrier (MeI_4), is accompanied by a thermal effect. The thermal dissociation of iodide results in the formation of a number of high-purity metals, for example, Ti, Zr, Ta, Nb, V, and others. As an example, table shows the temperatures (t_2) of settling of dissociated metals on the wire and temperatures (t_1) of synthesis and evaporation of the iodides of metals.

The iodide method is also used for the refining of metals to remove both interstitial admixtures, oxygen, nitrogen, carbon and also metallic admixtures which form iodides with lower volatility than the main metal. This process cannot be used in the case of metals whose melting point is lower than the dissociation temperature of the iodide. This would make it impossible to maintain these metals as a melt on the wire.

From the kinetic viewpoint, the process of iodide refining consists of the following stages:

- Synthesis of the iodide as a result of the reaction of the vapors of iodine with a metal;
- Transfer of the iodide to the heated wire;
- Thermal dissociation of iodide on the wire and deposition of the metal on the wire;

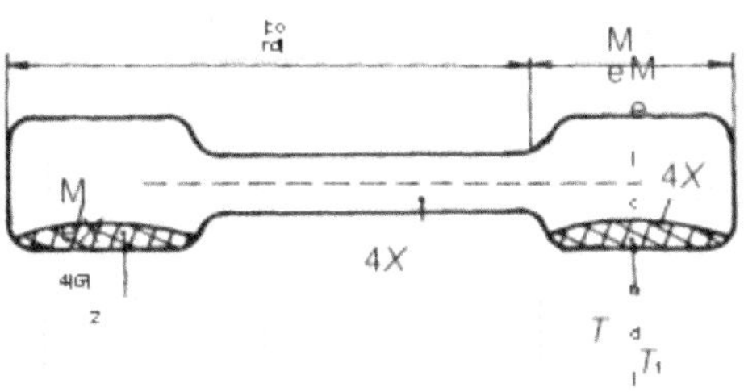

Figure: Diagram of distillation by synthesis and thermal dissociation of metal iodide $Me_{[cond]}+21_{2[G]}\underset{T_2}{\overset{T_1}{\rightleftarrows}}Mel_{4[G]}$

Table: Transport reaction of chlorides and subchlorides of metals

Element	Transport compound	Transport reaction	Conditions t_1/t_2 [°C]
		1. Transport of metals by means of sub-compounds	
Be	NaCl	$Be + 2NaCl = BeCl_2 + 2Na$	1000/21
B	BCl_3	$2B + BCl_3 = 3BCl$	
Ga	$GaCl_3$	$2Ga + GaCl_3 = 3GaCl$	
Zr	$ZrCl_4$	$Zr + 3ZrCl_4 = 4\ ZrCl_3$	
Nb	$NbCl_5$	$Nb + 4NbCl_5 = 5NbCl_5$	
Ta	$TaCl_5$	$Ta + 4TaCl_5 = 5TaCl_4$	
Ti	$TiCl_3$	$Ti + 2TiCl_3 = 3TiCl_2$	1200/1000^
Mo	Cl_2	$Mo + 5/2Cl_2 = MoCl_5$	1400/400
W	Cl_2	$W + 3Cl_2 = WCl_6$	1400/400
Au	Cl_2	$2Au + 3Cl_2 = Au_2Cl_6$	
Al	$AlCl_3$	$2Al + AlCl_3 = 3AlCl$	1000/600
Si	$SiCl_4$	$Si + SiCl_4 = 2SiCl_2$	
Si	$AlCl_3$	$Si + AlCl_3 = SiCl_2 + AlCl$	1000/21
Fe	HCl	$Fe + 2HCl = FeCl_2 + H_2$	1000/800
Co	HCl	$Co + 2HCl = CoCl_2 + H_2$	900/600
Cu	HCl	$3Cu + 2HCl = CuCl + 1/2H_2$	600/500

2. Transport of chlorides and oxychlorides of metals

VCl_3	Cl_2	$VCl_3 + 1/2Cl_2 = VCl_4$	300/250
$VOCl$	VCl_4	$VOCl + 2VCl_4 = VOCl_3 + 2VCl_3$	800/700
$NbCl_3$	$NbCl_5$	$NbCl_3 + NbCl_5 = 2NbCl_4$	390/355
$NbOCl_2$	$NbCl_5$	$NbOCl_2 + NbCl_5 = NbOCl_3 + NbCl_4$	400/350
$TaCl_3$	$TaCl_5$	$TaCl_3 + TaCl_5 = 2\ TaCl_4$	580/360
$TiOCl$	HCl	$TiOCl + 2HCl = TiCl_3 + H_2O$	650/550
$MoCl_3$	$MoCl_5$	$MoCl_3 + MoCl_5 = 2MoCl_4$	300/250
$AlOCl$	$NbCl_5$	$AlOCl + NbCl_5 = 1/2Al_2Cl_6 + NbOCl$	400/380
$CrOCl$	$CrCl_4+Cl_2$	$2CrOCl + CrCl_4 + Cl_2 = CrO_2Cl_2 + 2CrCl_3$	1000/840
$FeOCl$	HCl	$FeOCl + 2HCl = FeCl_3 + H_2O$	~350

3. Transport of oxides and other compounds

BeO	HCl	$BeO + 2HCl = BeCl_2 + H_2O$	$1000/800^{D}$
Ta_2O_5	Cl_2	$Ta_2O_5 + 3Cl_2 = 2TaOCl_3 + 3/2O_2$	$950/700^{A}$
Ta_2O_5	HCl	$Ta_2O_5 + 6HCl = 2TaOCl_3 + 3H_2O$	$1000/600^{D}$
Ta_2O_5	$TaCl_5$	$Ta_2O_5 + 3TaCl_5 = 5TaOCl_3$	$750/650^{A}$
Al_2O_3	HCl	$Al_2O_3 + 6HCl = 2AlCl_3 + 3H_2O$	$1000/21^{D}$
Al_2O3	Cl_2	$Al_2O_3 + 3Cl_2 = 2AlCl_3 + 3/2O_2$	$1240/21^{D}$
Fe_2O_3	HCl	$Fe_2O_3 + 6HCl = 2FeCl_3 + 3H_2O$	$1000/800^{D.A}$
SiO_2	$TaCl_5$	$SiO_2 + 2TaCl_5 = SiCl_4 + 2TaOCl_3$	1050/21
$FeWO_4$	HCl	$FeWo_4 + 6HCl = FeCl_2 + WOCl_4 + 3H_2O$	
MoS_2	Cl_2	$MoS_2 + 7/2Cl_2 = MoCl_5 + S_2Cl_2$	950/800
TiO_2	HCl	$TiO_2 + 4HCl_4 = TiCl_4 + 2H_2O$	1000/21

Table: Pure metals produced by thermal dissociation of their iodides

Element	Reaction temperature in vessel t_1 [° C]	Dissociation temperature on wire t_2 [°C]	Element	Reaction temperature in vessel t_1 [° C]	Dissociation temperature on wire t_2 [°C]
Be	400	800	Si	200	1000
Cr	800	1100	Ta	500	1100
Cu	400	1100	Ti	200	1400
Hf	800	1600	Cr	850	1150
Nb	360	900	V	900	1300
Fe	500	1100	Zr	200	1400

- Transfer of released iodine from the wire back to the iodine container. The kinetics of association of volatile iodide, i.e. the rate of deposition of metal, is affected by the following factors:
- Amount of iodine in the vessel;
- Reaction temperature in the vessel;
- Temperature of heated wire;
- Formation of lower iodides.

The metals which can be efficiently refined by the van Arkel method, in which all the previously mentioned reactions take place in the single reaction vessel, include Ti, Zr, Hf. At temperatures at which the refining process is realised by volatile iodides, various compounds of the metals to be refined, such as nitrides, carbonitrides, silicides and also other intermetallic compounds, such as oxides, carbides, do not take part in these iodide transport reactions. The iodides of the bivalent metals, if they form, are characterised by insignificant vapor pressure and, consequently, do not take part in the transfer process.

The iodide process can also be used for the production of alloys, if the reaction vessel is filled with a powder mixture of metals. For example, this method was used to produce alloys: Zr–Sn, Zr–Al, Ti- Zr, Zr–Hf, Zr–Cr, Zr–Ni, with a different concentration of the components.

Liquation Method

Liquation is a metallurgical process in common use starting around 600 years ago. It is a way to separate valuable metals from ores that are a mixture of two or more valuable metals, by a process of heating the ore until the metal with a lower melting point drains away. The purification of metals goes back to ancient times, with methods for gold purification traced back to 6,000 BC. The process of liquation does not stretch back that far, as it only works well with certain types of natural alloys, and is a specialized process.

Up until the 13th century, only seven metals were known to exist in nature: gold, silver, copper, mercury, tin, iron, and lead. Until the start of the 17th century, the metallurgical practice for separating metals from ore most commonly involved introducing either carbon or hydrogen compounds into the furnace. Germany began widespread use of the liquation practice in the 16th century to separate silver from copper.

The two earliest uses for liquation were the separation of silver from copper with lead as a solvent, and for removing tin from several types of minerals. In order for liquation to work, it must be done in the absence of air when lead is used, as the lead would not serve as a suitable solvent and would separate into litharge, otherwise, a form of earthy, poisonous lead solid with the chemical formula PbO. For this reason, liquation cannot be performed in an ordinary smelting furnace.

Metallurgical processes for liquation result initially in only partial separation of the alloys' metals. A typical copper-silver alloy may yield liquated lead that still contains 1-3% of copper, 10-30% lead, and the remainder as silver. The process is continued until sufficient silver is present in the liquated lead that is drained off, and this mixture is then cupelled, or refined further at the bottom of the furnace. An additional metallurgical method then takes place known as drying, essentially a prolongation of the liquation process to remove more lead from the remaining silver.

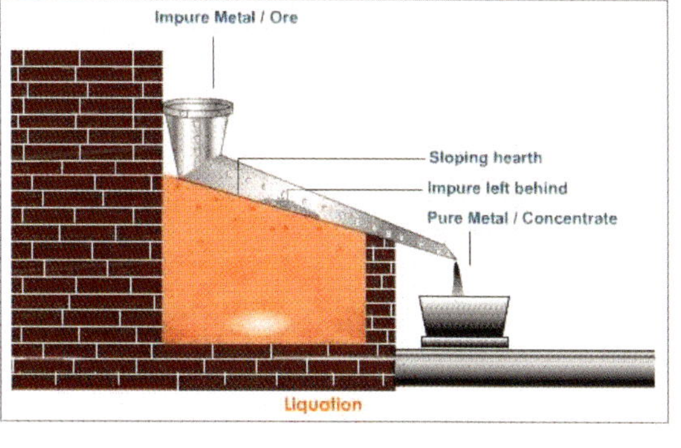

Though it seems simple and straightforward, the process of liquation is lengthy, requires special furnace conditions, and can yield ambiguous results as to the final composition percentages of metal. Metal is also lost in the process as slag, and, due to this and the length of processing, liquation has now been replaced by more efficient metallurgical methods. Nevertheless, liquation was so useful during the Renaissance period of history for its massive production of silver that it has been said to rival the invention of the printing press in importance, and is credited with reviving large parts of the European economy of the 15th-16th centuries.

Electrolytic Refining

Electrolysis, or electrolytic refining, is a technique used for extraction, as well as purification, of metals obtained by any refining method. In the electro-refining process, a block of impure metal becomes the anode, and a thin sheet of metal becomes the cathode. Both the cathode and anode are dipped in an electrolytic cell containing an aqueous solution of the metal salt. Upon the application of electric current of a suitable voltage, pure metal is deposited at the cathode by the dissolution of impure metal at the anode.

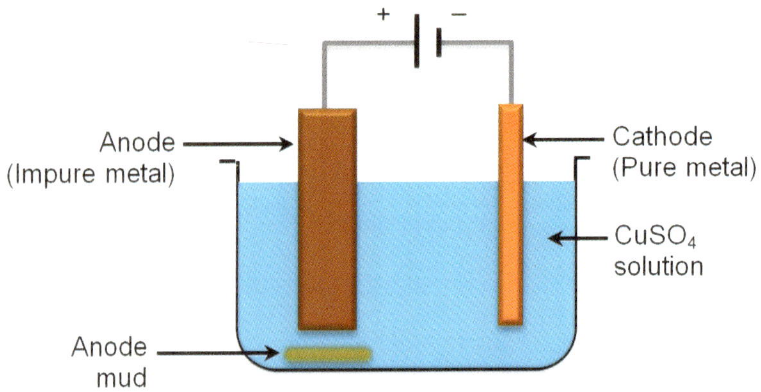

Usually the object of electrolytic refining is to separate one metal in pure form from an alloy containing a high percentage of the desired metal, copper for instance, and a number of other metals dissolved in the copper or intimately mixed with it. The impurities can be classified in two groups: (a) those metals which can be more easily oxidized than copper, e.g., iron, nickel, and (b) those metals less easily oxidized than copper —e.g., gold, silver.

Copper

In the actual copper-refining operation, the impure alloy of copper, nickel, gold, etc., is cast in a thin, at plate constituting the anode: the cathode is usually a sheet of pure electrolytic copper. Anodes and cathodes (30 to 40 of each) are immersed in a cell containing copper sulphate and sulphuric acid which serves as an electrolyte. The action of the electric current can be easily visualized of it is recalled that the action at the anode is always oxidizing and at the cathode reducing. The passage of the current tends to oxidize iron, nickel, and copper at the anode. No gold or silver will oxidize at the anode as long as there is metallic copper present. The metal oxides are soluble in the sulphuric acid of the electrolyte – and, as sulphates, are free to move to the cathode. At the cathode the copper sulphate, being more easily reduced than other sulphates present, is broken down to copper metal and sulphuric acid. The sulphuric acid returns to the anode to dissolve more copper, thus completing the cycle. None of the other sulphates in the electrolyte will be decomposed at the cathode as long as there is copper sulphate for the current to work on. In this way, the copper is separated by preferential solution from gold and silver which remain at the anode in metallic form so finely divided that they form a slime; it is separated from iron and nickel at the cathode by preferential decomposition of the sulphate.

The energy relationships in a cell of this type are characteristic of those in all cells where soluble anodes are used. When a compound such as copper sulphate is formed from copper, oxygen, and SO_3, a certain amount of energy is released or made available in the form of heat or equivalent electrical units. Conversely, when copper sulphate is decomposed to form copper, oxygen, and SO_3 exactly the reverse is true—an equal amount of energy (heat or electrical) is absorbed. In the cell with soluble anodes, both reactions are proceeding at the same time, so that energy required for decomposition is derived from the energy resulting from combination, the final result being that no energy is absorbed or dissipated. Hence, the only new electrical energy required to operate the soluble anode cell is represented by the voltage necessary to overcome polarization, cell resistance, etc. This voltage is relatively small in the case of copper sulphate—sulphuric acid electrolytes.

Other impure metals are refined in exactly the same way, though it is necessary to adjust the electrolytes and other operating conditions to suit each individual metal, the essential requirement being that the electrolyte must be composed of solution of a soluble salt of the metal to be refined.

Lead

A number of electrolytic lead plants are in operation, separating lead from copper, antimony, bismuth, gold, silver, etc. In this operation the anode is only 98 per cent lead, with antimony, copper, gold, silver, etc., present as impurities. A solution of lead fluosilicate and hydro-fluosilicic acid forms the electrolyte—sulphuric acid would not do, as lead sulphate is insoluble. The action in the cell is, with obvious modifications, identical with that in the copper refining cell, with the exception that any tin present in the anode is dissolved with the lead and deposited on the cathode.

Nickel

Nickel is refined electrolytically from metallic nickel or nickel sulphide anodes containing as impurities iron, copper, gold, silver, platinum, palladium etc. The electrolyte contains, principally, nickel sulphate in a nearly neutral solution. Extraordinary precautions must be taken to ensure that the electrolyte at the cathode is free of copper, iron, and other impurities. Each cathode is suspended in an individual porous- walled compartment to which pure electrolyte is fed at a predetermined rate. The impure electrolyte leaving the anodes is removed from the cell and is purified by removal of iron, copper, etc., before being returned to the cathode compartments for precipitation of pure nickel on the cathode. Otherwise, the basis of operations resembles that in copper refining.

Silver and Gold

Silver and gold are commonly refined electrolytically—a neutral silver nitrate solution is the electrolyte for silver refining, and gold chloride solution for gold refining.

Zone Refining

Zone melting (or zone refining or floating zone process or travelling melting zone) is a group of similar methods of purifying crystals, in which a narrow region of a crystal is melted, and this molten zone is moved along the crystal. The molten region melts impure solid at its forward edge and leaves a wake of purer material solidified behind it as it moves through the ingot. The impurities concentrate in the melt, and are moved to one end of the ingot. Zone refining was invented by John Desmond Bernal and further developed by William Gardner Pfann in Bell Labs as a method to prepare high purity materials, mainly semiconductors, for manufacturing transistors. Its early use was on germanium for this purpose, but it can be extended to virtually any solute-solvent system having an appreciable concentration difference between solid and liquid phases at equilibrium. This process is also known as the float zone process, particularly in semiconductor materials processing.

Process Details

The principle is that the segregation coefficient *k* (the ratio of an impurity in the solid phase to that in the liquid phase) is usually less than one. Therefore, at the solid/liquid boundary, the impurity atoms will diffuse to the liquid region. Thus, by passing a crystal boule through a thin section of furnace very slowly, such that only a small region of the boule is molten at any time, the impurities will be segregated at the end of the crystal. Because of the lack of impurities in the leftover regions which solidify, the boule can grow as a perfect single crystal if a seed crystal is placed at the base to initiate a chosen direction of crystal growth. When high purity is required, such as in semiconductor industry, the impure end of the boule is cut off, and the refining is repeated.

In zone refining, solutes are segregated at one end of the ingot in order to purify the remainder, or to concentrate the impurities. In zone leveling, the objective is to distribute solute evenly throughout the purified material, which may be sought in the form of a single crystal. For example, in the preparation of a transistor or diode semiconductor, an ingot of germanium is first purified by zone refining. Then a small amount of antimony is placed in the molten zone, which is passed through the pure germanium. With the proper choice of rate of heating and other variables, the antimony can be spread evenly through the germanium. This technique is also used for the preparation of silicon for use in computer chips.

Heaters

A variety of heaters can be used for zone melting, with their most important characteristic being the ability to form short molten zones that move slowly and uniformly through the ingot. Induction coils, ring-wound resistance heaters, or gas flames are common methods. Another method is to pass an electric current directly through the ingot while it is in a magnetic field, with the resulting magnetomotive force carefully set to be just equal to the weight in order to hold the liquid suspended. Optical heaters using high powered halogen or xenon lamps are used extensively in research facilities particularly for the production of insulators, but their use in industry is limited by the relatively low power of the lamps, which limits the size of crystals produced by this method. Zone melting can be done as a batch process, or it can be done continuously, with fresh impure material being continually added at one end and purer material being removed from the other, with impure zone melt being removed at whatever rate is dictated by the impurity of the feed stock.

Indirect-heating floating zone methods use an induction-heated tungsten ring to heat the ingot radiatively, and are useful when the ingot is of a high-resistivity semiconductor on which classical induction heating is ineffective.

Mathematical Expression of Impurity Concentration

When the liquid zone moves by a distance dx, the number of impurities in the liquid change. Impurities are incorporated in the melting liquid and freezing solid.

k_O : Segregation coefficient

L : Zone length

C_O : Initial uniform impurity concentration of the rod

C_L : Concentration of impurities in the liquid

I : Number of impurities in the liquid

I_O : Number of impurities in zone when first formed at bottom

The number of impurities in the liquid changes in accordance with thc expression below during the movement dx of the molten zonc

$$dI = (C_O - k_O C_L)dx$$

$$C_L = I / L$$

$$\int_0^x dx = \int_{I_O}^{I} \frac{dI}{C_O - \frac{k_O I}{L}}$$

$$I_O = C_O L$$

$$C_S = k_O I / L$$

$$C_S(x) = C_O\left(1-(1-k_O)e^{-\frac{k_O x}{L}}\right)$$

Applications

Solar Cells

In solar cells float zone processing is particularly useful because the single crystal silicon grown has desirable properties. The bulk charge carrier lifetime in float-zone silicon is the highest among various manufacturing processes. Float-zone carrier lifetimes are around 1000 microseconds compared to 20-200 microseconds with Czochralski process, and 1–30 microseconds with cast multi-crystalline silicon. A longer bulk lifetime increases the efficiency of solar cells significantly.

Related Processes

Zone Remelting

Another related process is zone remelting, in which two solutes are distributed through a pure metal. This is important in the manufacture of semiconductors, where two solutes of opposite conductivity type are used. For example, in germanium, pentavalent elements of group V such as antimony and arsenic produce negative (n-type) conduction and the trivalent elements of group III such as aluminum and boron produce positive (p-type) conduction. By melting a portion of such an ingot and slowly refreezing it, solutes in the molten region become distributed to form the desired n-p and p-n junctions.

Chromatographic Method

Chromatographic method follows the principles of chromatography. Chromatography deals with the movements of components at different rates in a mixture and/or differential absorption of an absorbent.

In this process, impure metal is put in a medium (liquid or a gas). Then we move the medium through an absorbent. Different components of the impure metal are will absorb at different levels. And then the components that were absorbed are removed by using a suitable solvent.

There are various forms of this method, like column chromatography (where we use Al_2O_3), thin layer chromatography, gas-liquid chromatography etc.

Types of Chromatographic Method

There are different kinds of chromatographic techniques and these are classified according to the shape of bed, physical state of mobile phase, separation mechanisms.

Apart from these there are certain modified forms of these chromatographic techniques involving different mechanisms and are hence categorized as modified or specialized chromatographic techniques.

Column Chromatography

It is the preparative application of chromatography. It is used to obtain pure chemical compounds from a mixture of compounds on a scale from micrograms up to kilograms using large industrial columns. The classical preparative chromatography column is a glass tube with a diameter from 5 to 50 mm and a height of 50 cm to 1 m with a tap at the bottom.

Slurry is prepared of the eluent with the stationary phase powder and then carefully poured into the column. Care must be taken to avoid air bubbles. A solution of the organic material is pipetted on top of the stationary phase.

This layer is usually topped with a small layer of sand or with cotton or glass wool to protect the shape of the organic layer from the velocity of newly added eluent. Eluent is slowly passed through the column to advance the organic material. Often a spherical eluent reservoir or an eluent-filled and stoppered separating funnel is put on top of the column.

The individual components are retained by the stationary phase differently and separate from each other while they are running at different speeds through the column with the eluent. At the end of the column they elute one at a time. During the entire chromatography process the eluent is collected in a series of fractions.

A chemist using column chromatography apparatus

The composition of the eluent flow can be monitored and each fraction is analyzed for dissolved compounds, e.g., by analytical chromatography, UV absorption, or fluorescence. Coloured compounds (or fluorescent compounds with the aid of an UV lamp) can be seen through the glass wall as moving bands.

The stationary phase or adsorbent in column chromatography is a solid. The most common stationary phase for column chromatography is $—C_{18}H_{37}$, followed by alumina. Cellulose powder has often been used in the past. Also possible are ion exchange chromatography, reversed-phase chromatography (RP), affinity chromatography or expanded bed adsorption (EBA). The stationary phases are usually finely ground powders or gels and/or are micro porous for an increased surface; though in EBA a fluidized bed is used.

The mobile phase or eluent is either a pure solvent or a mixture of different solvents. It is chosen so that the retention factor value of the compound of interest is roughly around 0.75 in order to minimize the time and the amount of eluent to run the chromatography. The eluent has also been chosen so that the different compounds can be separated effectively. The eluent is optimized in small scale pretests, often using thin layer chromatography (TLC) with the same stationary phase.

A faster flow rate of the eluent minimizes the time required to run a column and thereby minimizes diffusion, resulting in a better separation. A simple laboratory column runs by gravity flow. The flow rate of such a column can be increased by extending the fresh eluent filled column above the top of the stationary phase or decreased by the tap controls. Better flow rates can be achieved by using a pump or by using compressed gas (e.g., air, nitrogen, or argon) to push the solvent through the column (flash column chromatography).

Automated flash chromatography systems attempt to minimize human involvement in the purification process. Automated systems may include components normally found on HPLC systems (gradient pump, sample injection apparatus, UV detector) and a fraction collector to collect the eluent. The software controlling an automated system will coordinate the components and help the user to find the resulting purified material within the fraction collector. The software will also store results from the process for archival or later recall purposes.

Paper Chromatography

It is an analytical technique for separating and identifying mixtures that are or can be coloured, especially pigments. This can also be used in secondary or primary schools in ink experiments. This method has been largely replaced by thin layer chromatography; however it is still a powerful teaching tool. Two-way paper chromatography, also called two-dimensional chromatography, involves using two solvents and rotating the paper 90° in between. This is useful for separating complex mixtures of similar compounds, for example, amino acids.

A small, ideally concentrated spot of solution that contains the sample is applied to a strip of chromatography paper about 1 cm from the base, usually using a capillary tube for maximum precision. This sample is absorbed onto the paper and may form interactions with it. Any substance that reacts or bonds with the paper cannot be measured using -Solvent front technique. The paper is then dipped into a suitable solvent, such as ethanol or water, taking care that the spot is above the surface of the solvent, and placed in a sealed container.

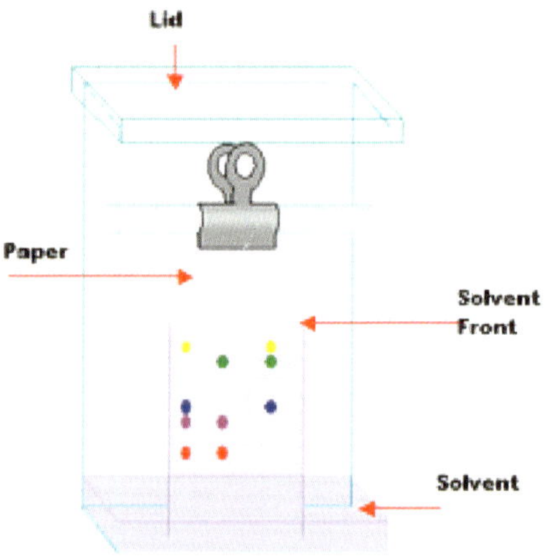

Schematic diagram of paper chromatography

The solvent moves up the paper by capillary action, which occurs as a result of the attraction of the solvent molecules to the paper and to one another. As the solvent rises through the paper it meets and dissolves the sample mixture, which will then travel up the paper with the solvent. Different compounds in the sample mixture travel at different rates due to differences in solubility in the solvent, and due to differences in their attraction to the fibers in the paper. Paper chromatography takes anywhere from several minutes to several hours.

In some cases, paper chromatography does not separate pigments completely; this occurs when two substances appear to have the same values in a particular solvent. In these cases, two-way chromatography is used to separate the multiple-pigment spots. The chromatogram is turned by ninety degrees, and placed in a different solvent in the same way as before; some spots separate in the presence of more than one pigment.

As before, the value is calculated, and the two pigments are identified. The R_f value (retention factor) is the distance travelled by a particular component from the origin (where the sample was originally spotted) as a ratio to the distance travelled by the solvent front from the origin. R_f values for each substance will be unique, and can be used to identify components. A particular component will have the same R_f value if it is separated under identical conditions.

After development, the spots corresponding to different compounds may be located by their colour, ultraviolet light, ninhydrin (Triketohydrindane hydrate) or by treatment with iodine vapours. The final chromatogram can be compared with other known mixture chromatograms to identify sample mixture using the R_n value.

As in most other forms of chromatography, paper chromatography uses R_n values to help identify compounds. R_f values are calculated by dividing the distance the pigment travels up the paper by the distance the solvent travels (the solvent front). Because R_f values are standard for a given compound, known R_n values can be used to aid in the identification of an unknown substance in an experiment.

Thin Layer Chromatography

Thin-layer chromatography (TLC) is a chromatographic technique that is useful for separating organic compounds. It involves a stationary phase consisting of a thin layer of adsorbent material, usually silica gel, aluminium oxide, or cellulose immobilized onto a flat, inert carrier sheet. A liquid phase consisting of the solution to be separated dissolved in an appropriate solvent is drawn through the plate via capillary action, separating the experimental solution.

When the solvent front reaches the other edge of the stationary phase, the plate is removed from the solvent reservoir. The separated spots are visualized with ultraviolet light or by placing the plate in iodine vapour. The different components in the mixture move up the plate at different rates due to differences in their portioning behavior between the mobile liquid phase and the stationary phase.

It can be used to determine the pigments a plant contains, to detect pesticides or insecticides in food, in forensics to analyze the dye composition of fibers, or to identify compounds present in a given substance, among other uses. It is a quick, generic method for organic reaction monitoring.

TLC plates are made by mixing the adsorbent, such as silica gel, with a small amount of inert binder like calcium sulphate (gypsum) and water. This mixture is spread as thick slurry on an unreactive carrier sheet, usually glass, thick aluminum foil, or plastic, and the resultant plate is dried and activated by heating in an oven for thirty minutes at 110° C. The thickness of the adsorbent layer is typically around 0.1-0.25 mm for analytical purposes and around 1-2 mm for preparative TLC. Every type of chromatography contains a mobile phase and a stationary phase.

The process is similar to paper chromatography with the advantage of faster runs, better separations, and the choice between different stationary phases. Because of its simplicity and speed TLC is often used for monitoring chemical reactions and for the qualitative analysis of reaction products.

A small spot of solution containing the sample is applied to a plate, about one centimetre from the base. The plate is then dipped into a suitable solvent, such as ethanol or water, and placed in a sealed container. The solvent moves up the plate by capillary action and meets the sample mixture, which is dissolved and is carried up the plate by the solvent.

Different compounds in the sample mixture travel at different rates due to differences in solubility in the solvent, and due to differences in their attraction to the stationary phase. Results also vary depending on the solvent used. For example, if the solvent were a 90:10 mixture of hexane to ethyl acetate, then the solvent would be mostly non-polar.

This means that when analyzing the TLC, the non-polar parts will have moved further up the plate. The polar compounds, in contrast, will not have moved as much. The reverse is true when using a solvent that is more polar than non-polar (10:90 hexane to ethyl acetate). With these solvents, the polar compounds will move higher up the plate, while the non-polar compounds will not move as much.

This electrically refined copper is 100% pure

The appropriate solvent in context of thin layer chromatography will be one which differs from the stationary phase material in polarity. If polar solvent is used to dissolve the sample and spot is applied over polar stationary phase of TLC, the sample spot will grow radially due to capillary action, which is not advisable as one spot may mix with the other.

Hence, to restrict the radial growth of sample-spot, the solvent used for dissolving samples in order to apply them on plates should be as non-polar or semi-polar as possible when the stationary phase is polar, and vice versa.

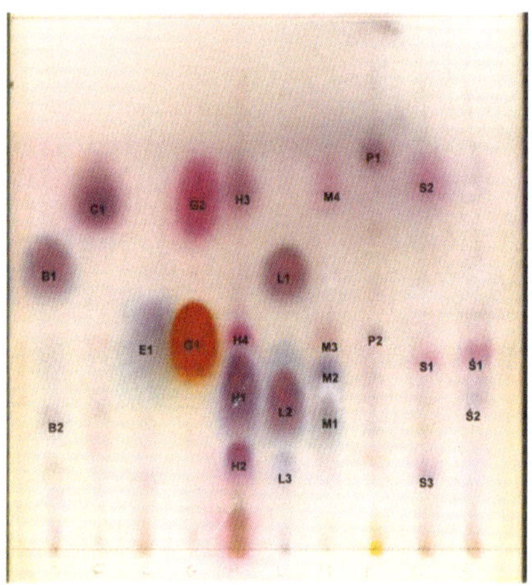

chromatogram of 10 essential oils coloured with vanillin reagent

As the chemicals being separated may be colourless, several methods exist to visualize the spots:

1. Often a small amount of a fluorescent compound, usually Manganese-activated Zinc Silicate, is added to the adsorbent that allows the visualization of spots under a black-light(UV_{254}).

The adsorbent layer will thus fluoresce light green by itself, but spots of analyte quench this fluorescence.

2. Iodine vapours are a general unspecific colour reagent.
3. Specific colour reagents exist into which the TLC plate is dipped or which are sprayed onto the plate.

Once visible, the R_f value of each spot can be determined by dividing the distance travelled by the product by the total distance travelled by the solvent (the solvent front). These values depend on the solvent used, and the type of TLC plate, and are not physical constants.

Applications

In organic chemistry, reactions are qualitatively monitored with TLC. Spots sampled with a capillary tube are placed on the plate: a spot of starting material, a spot from the reaction mixture, and a "co-spot" with both. A small (3 by 7 cm) TLC plate takes a couple of minutes to run.

The analysis is qualitative, and it will show if starting material has disappeared, product has appeared, and how many products are generated. Unfortunately, TLCs from low-temperature reactions may give misleading results, because the sample is warmed to room temperature in the capillary. One such reaction is DIBALH reduction of ester to aldehyde.

Gas-Liquid Chromatography (GLC) or Simply Gas Chromatography (GC)

It is a type of chromatography in which the mobile phase is a carrier gas, usually an inert gas such as helium or an unreactive gas such as nitrogen, and the stationary phase is a microscopic layer of liquid or polymer on an inert solid support, inside glass or metal tubing, called a column. The instrument used to perform gas chromatographic separations is called a gas chromatograph (also: aerograph, gas separator).

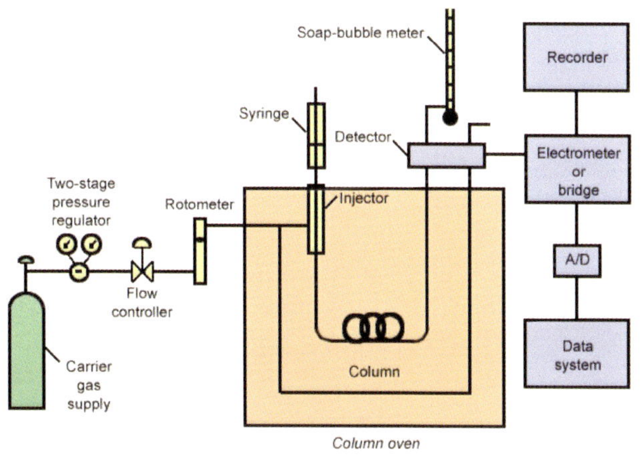

A gas chromatograph

A gas chromatograph is a chemical analysis instrument for separating chemicals in a complex sample. A gas chromatograph uses a flow-through narrow tube known as the column, through which different chemical constituents of a sample pass in a gas stream (carrier gas, mobile phase) at different rates depending on their various chemical and physical properties and their interaction with a specific column filling, called the stationary phase.

As the chemicals exit the end of the column, they are detected and identified electronically. The function of the stationary phase in the column is to separate different components, causing each one to exit the column at a different time (retention time). Other parameters that can be used to alter the order or time of retention are the carrier gas flow rate, and the temperature.

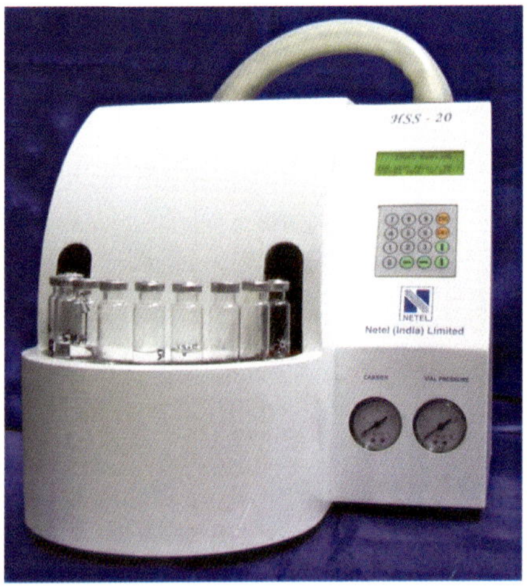

A gas chromatograph with an headspace sampler

In a GC analysis, a known volume of gaseous or liquid analyte is injected into the "entrance" (head) of the column, usually using a micro syringe (or, solid phase micro-extraction fibers, or a gas source switching system). As the carrier gas sweeps the analyte molecules through the column, this motion is inhibited by the adsorption of the analyte molecules either onto the column walls or onto packing materials in the column.

The rate at which the molecules progress along the column depends on the strength of adsorption, which in turn depends on the type of molecule and on the stationary phase materials. Since each type of molecule has a different rate of progression, the various components of the analyte mixture are separated as they progress along the column and reach the end of the column at different times (retention time).

A detector is used to monitor the outlet stream from the column; thus, the time at which each component reaches the outlet and the amount of that component can be determined. Generally, substances are identified (qualitatively) by the order in which they emerge (elute) from the column and by the retention time of the analyte in the column.

Auto Samplers

The auto sampler provides the means to introduce automatically a sample into the inlets. Manual insertion of the sample is possible but very rare nowadays. Automatic insertion provides better reproducibility and time-optimization. Different kinds of auto samplers exist. Auto samplers can be classified in relation to sample capacity (auto-injectors VS. auto samplers, where auto-injectors can work a small number of samples), to robotic technologies (XYZ robot VS rotating/SCARA-robot – the most common), or to analysis:

1. Liquid
2. Static head-space by syringe technology
3. Dynamic head-space by transfer-line technology
4. Solid phase micro extraction (SPME).

Inlets

The column inlet (or injector) provides the means to introduce a sample into a continuous flow of carrier gas. The inlet is a piece of hardware attached to the column head.

Common inlet types are:

- S/SL (Split/Split less) injector

  A sample is introduced into a heated small chamber via a syringe through a septum; the heat facilitates volatilization of the sample and sample matrix. The carrier gas then either sweeps the entirety (split less mode) or a portion (split mode) of the sample into the column. In split mode, a part of the sample/carrier gas mixture in the injection chamber is exhausted through the split vent.

- On-column inlet

  The sample is here introduced in its entirety without heat.

- PTV injector

  Temperature-programmed sample introduction was first described by Vogt in 1979. Originally Vogt developed the technique as a method for the introduction of large sample volumes (up to 250 μl) in capillary GC. Vogt introduced the sample into the liner at a controlled injection rate. The temperature of the liner was chosen slightly below the boiling point of the solvent.

  The low-boiling solvent was continuously evaporated and vented through the split line. Based on this technique, Poy developed the Programmed Temperature Vaporizing Injector PTV. By introducing the sample at a low initial liner temperature many of the disadvantages of the classic hot injection techniques could be circumvented.

- Gas source inlet or gas switching valve

  Gaseous samples in collection bottles are connected to what is most commonly a six-port switching valve. The carrier gas flow is not interrupted while a sample can be expanded into a previously evacuated sample loop. Upon switching, the contents of the sample loop are inserted into the carrier gas stream.

- P/T (Purge-and-Trap) system

  An inert gas is bubbled through an aqueous sample causing insoluble volatile chemicals to be purged from the matrix. The volatiles are 'trapped' on an absorbent column (known as a trap or concentrator) at ambient temperature. The trap is then heated and the volatiles are directed into the carrier gas stream. Samples requiring pre-concentration or purification can be introduced via such a system, usually hooked up to the S/SL port.

- SPME (solid phase micro extraction) offers a convenient, low-cost alternative to P/T systems with the versatility of a syringe and simple use of the S/SL port.

Columns

Two Types of Columns are used in GC

a. Packed columns are 1.5-10 m in length and have an internal diameter of 2-4 mm. The tubing is usually made of stainless steel or glass and contains a packing of finely divided, inert, solid support material (e.g., diatomaceous earth) that is coated with a liquid or solid stationary phase. The nature of the coating material determines what type of materials will be most strongly adsorbed. Thus numerous columns are available that are designed to separate specific types of compounds.

b. Capillary columns have a very small internal diameter, on the order of a few tenths of millimeters, and lengths between 25-60 metres are common. The inner column walls are coated with the active materials (WCOT columns), some columns are quasi-solid filled with many parallel microspores (PLOT columns). Most capillary columns are made of fused-silica with a polyimide outer coating. These columns are flexible, so a very long column can be wound into a small coil.

c. New developments are sought where stationary phase incompatibilities lead to geometric solutions of parallel columns within one column.

Among these new developments are:

1. Internally heated micro FAST columns, where two columns, an internal heating wire and a temperature sensor are combined within a common column sheath (micro FAST);

2. Micro packed columns (1/16″ OD) are column-in-column packed columns where the outer column space has a packing different from the inner column space, thus providing the separation behaviour of two columns in one. They can be easily fit to inlets and detectors of a capillary column instrument.

The temperature-dependence of molecular adsorption and of the rate of progression along the column necessitates a careful control of the column temperature to within a few tenths of a degree for precise work. Reducing the temperature produces the greatest level of separation, but can result in very long elution times. For some cases, temperature is ramped either continuously or in steps to provide the desired separation. This is referred to as a temperature program. Electronic pressure control can also be used to modify flow rate during the analysis, aiding in faster run times while keeping acceptable levels of separation.

The choice of carrier gas (mobile phase) is important, with hydrogen being the most efficient and providing the best separation. However, helium has a larger range of flow rates that are comparable to hydrogen in efficiency, with the added advantage that helium is non-flammable, and works with a greater number of detectors. Therefore, helium is the most common carrier gas used.

Detectors

A number of detectors are used in gas chromatography. The most common are the flame ionization detector (FID) and the thermal conductivity detector (TCD). Both are sensitive to a wide range of components, and both work over a wide range of concentrations.

While TCDs are essentially universal and can be used to detect any component other than the carrier gas (as long as their thermal conductivities are different than that of the carrier gas, at detector temperature), FIDs are sensitive primarily to hydrocarbons, and are more sensitive to them than TCD.

However, an FID cannot detect water. Both detectors are also quite robust. Since TCD is non-destructive, it can be operated in series before an FID (destructive), thus providing complementary detection of the same eluents. Other detectors are sensitive only to specific types of substances, or work well only in narrower ranges of concentrations.

They include:

1. Discharge ionization detector (DID)
2. Electron capture detector (ECD)
3. Flame photometric detector (FPD)
4. Hall electrolytic conductivity detector (E1CD)
5. Helium ionization detector (HID)
6. Nitrogen phosphorus detector (NPD)
7. Mass selective detector (MSD)
8. Photo-ionization detector (PID)
9. Pulsed discharge ionization detector (PDD)

Some gas chromatographs are connected to a mass spectrometer which acts as the detector. The combination is known as GC-MS. Some GC-MS are connected to an Nuclear Magnetic Resonance Spectrometer which acts as a back-up detector. This combination is known as GC-MS- NMR. Some GC-MS-NMR are connected to an infrared spectra which acts as a back-up detector. This combination is known as GC-MS-NMR-IR. It must, however, be stressed that this is very rare as most analysis needed can be concluded via purely GC-MS.

Methods

The method is the collection of conditions in which the GC operates for a given analysis. Method development is the process of determining what conditions are adequate and/or ideal for the analysis required. Conditions which can be varied to accommodate a required analysis include inlet temperature, detector temperature, column temperature and temperature program, carrier gas and carrier gas flow rates, the column's stationary phase, diameter and length, inlet type and flow rates, sample size and injection technique.

Depending on the detector(s) installed on the GC, there may be a number of detector conditions that can also be varied. Some GCs also include valves which can change the route of sample and carrier flow, and the timing of the turning of these valves can be important to method development.

Carrier Gas Selection and Flow Rates

Typical carrier gases include helium, nitrogen, argon, hydrogen and air. Which gas to use is usually determined by the detector being used, for example, a DID requires helium as the carrier gas. When analyzing gas samples, however, the carrier is sometimes selected based on the sample's matrix, for example, when analyzing a mixture in argon, an argon carrier is preferred, because the argon in the sample does not show upon the chromatogram. Safety and availability can also influence carrier selection, for example, hydrogen is flammable, and high-purity helium can be difficult to obtain in some areas of the world.

The purity of the carrier gas is also frequently determined by the detector, though the level of sensitivity needed can also play a significant role. Typically, purities of 99.995% or higher are used. Trade names for typical purities include "Zero Grade", "Ultra-High Purity (UHP) Grade", "4.5 Grade" and "5.0 Grade".

The carrier gas flow rate affects the analysis in the same way that temperature does. The higher the flow rates the faster the analysis, but the lower the separation between analytes. Selecting the flow rate is, therefore, the same compromise between the level of separation and length of analysis as selecting the column temperature.

With GCs made before 1990s, carrier flow rate was controlled indirectly by controlling the carrier inlet pressure, or "column head pressure". The actual flow rate was measured at the outlet of the column or the detector with an electronic flow meter, or a bubble flow meter, and could be an involved, time consuming, and frustrating process. The pressure setting was not able to be varied during the run, and thus the flow was essentially constant during the analysis.

Many modern GCs, however, electronically measure the flow rate, and electronically control the carrier gas pressure to set the flow rate. Consequently, carrier pressures and flow rates can be adjusted during the run, creating pressure/flow programs similar to temperature programs.

Inlet Types and Flow Rates

The choice of inlet type and injection technique depends on if the sample is in liquid, gas, adsorbed, or solid form, and on whether a solvent matrix is present that has to be vaporized. Dissolved samples can be introduced directly onto the column via a COC injector, if the conditions are well known; if a solvent matrix has to be vaporized and partially removed, a S/SL injector is used (most common injection technique); gaseous samples (e.g., air cylinders) are usually injected using a gas switching valve system; adsorbed samples (e.g., on adsorbent tubes) are introduced using either an external (on-line or off-line) desorption apparatus such as a purge-and-trap system, or are desorbed in the S/SL injector (SPME applications).

Sample Size and Injection Technique

The real chromatographic analysis starts with the introduction of the sample onto the column. The development of capillary gas chromatography resulted in many practical problems with the injection technique. The technique of on-column injection, often used with packed columns, is usually not possible with capillary columns.

The injection system, in the capillary gas chromatograph, should fulfil the following two requirements:

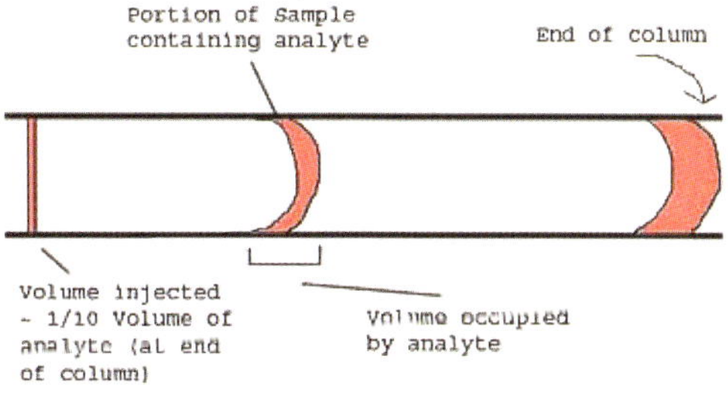

The rule of ten in gas chromatography

1. The amount injected should not overload the column.
2. The width of the injected plug should be small compared to the spreading due to the chromatographic process. Failure to comply with this requirement will reduce the separation capability of the column. As a general rule, the volume injected, V_{inj}, and the volume of the detecor cell, V_{det}, should be about 1/10 of the volume occupied by the portion of sample containing the molecules of interest (analytes) when they exit the column.

Some general requirements, which a good injection technique should fulfil, are:

1. It should be possible to obtain the column's optimum separation efficiency.
2. It should allow accurate and reproducible injections of small amounts of representative samples.
3. It should induce no change in sample composition. It should not exhibit discrimination based on differences in boiling point, polarity, concentration or thermal/catalytic stability.
4. It should be applicable for trace analysis as well as for undiluted samples.

A gas chromatography oven open to show a capillary column

Column Selection

The column in a GC is contained in an oven, the temperature of which is precisely controlled electronically. The rate at which a sample passes through the column is directly proportional to the temperature of the column. The higher the column temperature, the faster the sample moves through the column. However, the faster a sample moves through the column, the less it interacts with the stationary phase, and the less the analytes are separated. In general, the column temperature is selected to compromise between the length of the analysis and the level of separation.

A method which holds the column at the same temperature for the entire analysis is called "isothermal". Most methods, however, increase the column temperature during the analysis, the initial temperature, rate of temperature increase (the temperature "ramp") and final temperature is called the "temperature program". A temperature program allows analytes that elute early in the analysis to separate adequately, while shortening the time it takes for late-eluting analytes to pass through the column.

Data Reduction and Analysis

Qualitative Analysis

Generally chromatographic data is presented as a graph of detector response (y-axis) against retention time (x-axis). This provides a spectrum of peaks for a sample representing the analytes present in a sample eluting from the column at different times. Retention time can be used to identify analytes if the method conditions are constant.

Also, the pattern of peaks will be constant for a sample under constant conditions and can identify complex mixtures of analytes. In most modern applications, however, the GC is connected to a mass spectrometer or similar detector that is capable of identifying the analytes represented by the peaks.

Quantitive Analysis

The area under a peak is proportional to the amount of analyte present. By calculating the area of the peak using the mathematical function of integration, the concentration of an analyte in the original sample can be determined. Concentration can be calculated using a calibration curve created by finding the response for a series of concentrations of analyte, or by determining the response factor of an analyte.

The response factor is the expected ratio of an analyte to an internal standard and is calculated by finding the response of a known amount of analyte and a constant amount of internal standard (a chemical added to the sample at a constant concentration, with a distinct retention time to the analyte). In most modern GC-MS systems, computer software is used to draw and integrate peaks, and match MS spectra to library spectra.

Application

In general, substances that vaporize below 300° C (and therefore are stable up to that temperature) can be measured quantitatively. The samples are also required to be salt-free; they should not contain ions. Very minute amounts of a substance can be measured, but it is often required that the sample must be measured in comparison to a sample containing the pure, suspected substance. Various temperature programs can be used to make the readings more meaningful; for example, to differentiate between substances that behave similarly during the GC process.

Professionals working with GC analyze the content of a chemical product; for example, in assuring the quality of products in the chemical industry; or measuring toxic substances in soil, air or water. GC is very accurate if used properly and can measure Pico moles of a substance in a 1 ml liquid sample, or parts-per-billion concentrations in gaseous samples.

In practical courses at colleges, students sometimes get acquainted to the GC by studying the contents of Lavender oil or measuring the ethylene that is secreted by Nicotiana benthamiana plants after artificially injuring their leaves. These GC analyses hydrocarbons (C2-C40+). In a typical experiment, a packed column is used to separate the light gases, which are then detected with a TCD.

The hydrocarbons are separated using a capillary column and detected with an FID. A complication with light gas analyses that include H_2 is that He, which is the most common and most sensitive inert carrier (sensitivity is proportional to molecular mass) has an almost identical thermal conductivity to hydrogen (it is the difference in thermal conductivity between two separate filaments in a Wheatstone Bridge type arrangement that shows when a component has been eluted).

For this reason, dual TCD instruments are used with a separate channel for hydrogen that uses nitrogen as a carrier. Argon is often used when analysing gas phase chemistry reactions such as F-T synthesis so that a single carrier gas can be used rather than 2 separate ones. The sensitivity is less but this is a trade-off for simplicity in the gas supply.

Limitations Associated with Gas Chromatography

In fact, a GC analysis takes much more time; sometimes a single sample must be run more than an hour according to the chosen program; and even more time is needed to "heat out" the column; so it is free from the first sample and can be used for the next. Equally, several runs are needed to confirm the results of a study—a GC analysis of a single sample may simply yield a result per chance.

Also, GC does not positively identify most samples; and not all substances in a sample will necessarily be detected. All a GC truly tells is at which relative time a component eluted from the column and that the detector was sensitive to it.

To make results meaningful, analysts need to know which components at which concentrations are to be expected; and even then a small amount of a substance can hide itself behind a substance having both a higher concentration and the same relative elution time. Last but not least it is often needed to check the results of the sample against a GC analysis of a reference sample containing only the suspected substance. A GC-MS can remove much of this ambiguity, since the mass spectrometer will identify the component's molecular weight. But this still takes time and skill to do properly.

Similarly, most GC analyses are not push-button operations. According to the substances one expects to find the operating program must be carefully chosen. A push-button operation can exist for running similar samples repeatedly, such as in a chemical production environment or for comparing 20 samples from the same experiment to calculate the mean content of the same substance. However, for the kind of investigative work this is clearly not the case.

High Performance Liquid Chromatography

High-performance liquid chromatography (HPLC) is a form of column chromatography used frequently in biochemistry and analytical chemistry. It is also sometimes referred to as high-pressure liquid chromatography. HPLC is used to separate components of a mixture by using a variety of chemical interactions between the substance being analyzed (analyte) and the chromatography column.

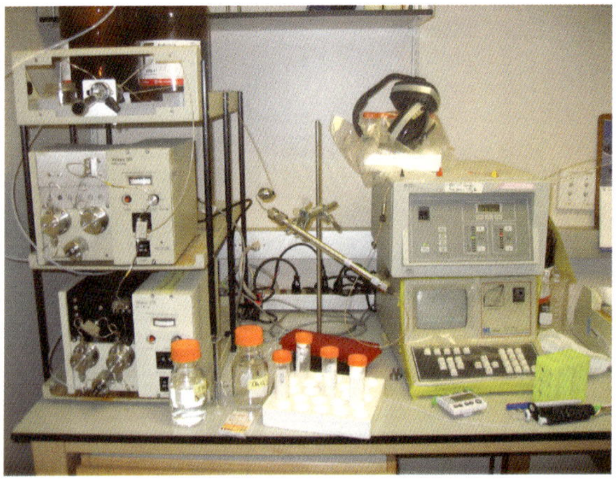

In isocratic HPLC, the analyte is forced through a column of the stationary phase (usually a tube packed with small round particles with a certain surface chemistry) by pumping a liquid (mobile phase) at high pressure through the column. The sample to be analyzed is introduced in a small volume to the stream of mobile phase and is retarded by specific chemical or physical interactions with the stationary phase as it traverses the length of the column.

The amount of retardation depends on the nature of the analyte, stationary phase and mobile phase composition. The time at which a specific analyte elutes (comes out of the end of the column) is called the retention time and is considered a reasonably unique identifying characteristic of a given analyte. The use of pressure increases the linear velocity (speed) giving the components less time to diffuse within the column, leading to improved resolution in the resulting chromatogram.

Common solvents used include any miscible combinations of water or various organic liquids (the most common are methanol and acetonitrile). Water may contain buffers or salts to assist in the separation of the analyte components, or compounds such as Trifluoroacetic acid which acts as an ion pairing agent.

A further refinement to HPLC has been to vary the mobile phase composition during the analysis; this is known as gradient elution. A normal gradient for reverse phase chromatography might start at 5% methanol and progress linearly to 50% methanol over 25 minutes, depending on how hydrophobic the analyte is.

The gradient separates the analyte mixtures as a function of the affinity of the analyte for the current mobile phase composition relative to the stationary phase. This partitioning process is similar to that which occurs during a liquid-liquid extraction but is continuous, not step-wise. In this example, using a water/methanol gradient, the more hydrophobic components will elute (come off the column) under conditions of relatively high methanol; whereas the more hydrophilic compounds will elute under conditions of relatively low methanol.

The choice of solvents, additives and gradient, depends on the nature of the stationary phase and the analyte. Often a series of tests are performed on the analyte and a number of generic runs may be processed in order to find the optimum HPLC method for the analyte — the method which gives the best separation of peaks.

Types of HPLC

Normal Phase Chromatography

Normal phase HPLC (NP-HPLC) was the first kind of HPLC chemistry used, and separates analytes based on polarity. This method uses a polar stationary phase and a non-polar mobile phase, and is used when the analyte of interest is fairly polar in nature. The polar analyte associates with and is retained by the polar stationary phase.

Adsorption strengths increase with increase in analyte polarity, and the interaction between the polar analyte and the polar stationary phase (relative to the mobile phase) increases the elution time. The interaction strength not only depends on the functional groups in the analyte molecule, but also on steric factors, and structural isomers are often resolved from one another.

Use of more polar solvents in the mobile phase will decrease the retention time of the analytes while more hydrophobic solvents tend to increase retention times. Particularly polar solvents in a mixture tend to deactivate the column by occupying the stationary phase surface.

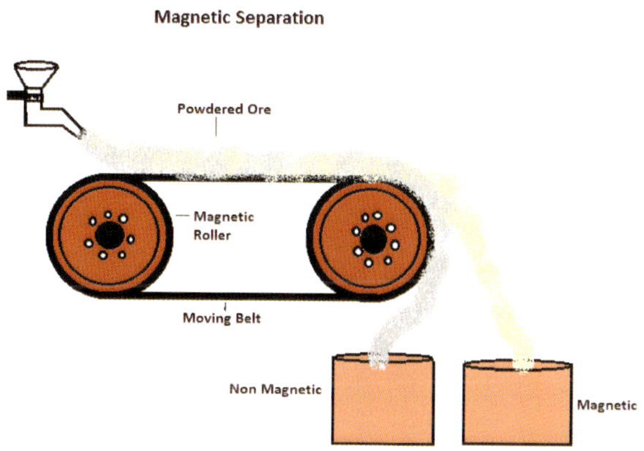

This is somewhat particular to normal phase because it is most purely an adsorptive mechanism (the interactions are with a hard surface rather than a soft layer on a surface).

NP-HPLC had fallen out of favour in the 1970s with the development of reversed-phase HPLC because of a lack of reproducibility of retention times as water or protic organic solvents changed the hydration state of the silica or alumina chromatographic media. Recently it has become useful again with the development of HILIC bonded phases which utilize a partition mechanism which provides reproducibility.

technique is generally combined with others that further separate molecules by other characteristics, such as acidity, basicity, charge, and affinity for certain compounds.

The underlying principle of SEC is that particles of different sizes will elute (filter) through a stationary phase at different rates. This result in the separation of a solution of particles based on size, provided that all the particles are loaded simultaneously or near simultaneously, particles of the same size should elute together.

This is usually achieved with an apparatus called a column, which consists of a hollow tube tightly packed with extremely small porous polymer beads designed to have pores of different sizes. These pores may be depressions on the surface or channels through the bead. As the solution travels down the column some particles enter into the pores. Larger particles cannot enter into as many pores. The larger the particles, the less overall volume to traverse over the length of the column, and the faster the elution.

The filtered solution that is collected at the end is known as the eluent. The void volume consists of any particles too large to enter the medium, and the solvent volume is known as the column volume. In real life situations, particles in solution do not have a constant, fixed size, resulting in the probability that a particle which would otherwise be hampered by a pore may pass right by it. Also, the stationary phase particles are not ideally defined; both particles and pores may vary in size.

Elution curves, therefore, resemble Gaussian distributions. The stationary phase may also interact in undesirable ways with a particle and influence retention times, though great care is taken by column manufacturers to use stationary phases which are inert and minimize this issue.

Like other forms of chromatography, increasing the column length will tighten the resolution, and increasing the column diameter increases the capacity of the column. Proper column packing is important to maximize resolution: an over packed column can collapse the pores in the beads, resulting in a loss of resolution. An under packed column can reduce the relative surface area of the stationary phase accessible to smaller species, resulting in those species spending less time trapped in pores.

In simple manual columns the eluent is collected in constant volumes, known as fractions. The more similar the particles are in size, the more likely they will be in the same fraction and not detected separately. More advanced columns overcome this problem by constantly monitoring the eluent.

The collected fractions are often examined by spectroscopic techniques to determine the concentration of the particles eluted. Three common spectroscopy detection techniques are refractive index (RI), evaporative light scattering (ELS), and ultraviolet (UV). When eluting spectroscopically similar species (such as during biological purification) other techniques may be necessary to identify the contents of each fraction.

The elution volume decreases roughly linearly with the logarithm of the molecular hydrodynamic volume (often assumed to be proportional to molecular weight). Columns are often calibrated using 4-5 standard samples (e.g., folded proteins of known molecular weight) to determine the void volume and the slope of the logarithmic dependence. This calibration may need to be repeated under different solution conditions.

Applications

Proteomics

SEC is generally considered a low resolution chromatography as it does not discern similar species very well, and is, therefore, often reserved for the final "polishing" step of a purification. The technique can determine the quaternary structure of purified proteins which have slow exchange times, since it can be carried out under native solution conditions, preserving macromolecular interactions.

SEC can also assay protein tertiary structure as it measures the hydrodynamic volume (not molecular weight), allowing folded and unfolded versions of the same protein to be distinguished. For example, the apparent hydrodynamic radius of a typical protein domain might be 14a and 36A for the folded and unfolded forms respectively.

SEC allows the separation of these two forms as the folded form will elute much later due to its smaller size. Alternatively, folded and unfolded versions of the same metalloproteinase can be separated according to their different isoelectric points by using quantitative preparative native continuous polyacrylamide gel electrophoresis (QPNC-PAGE).

Polymer Synthesis

SEC can be used as a measure of both the size and the polydispersity of a synthesised polymer; that one is able to find distribution of sizes of polymer molecules. If standards of a known size are run previously, then a calibration curve can be created to determine the sizes of polymer molecules of interest.

Alternatively, techniques such as light scattering and/or viscometry can be used online with SEC to yield absolute molecular weights that do not rely on calibration with standards of known molecular weight. Due to the difference in size of two polymers with identical molecular weights, the absolute determination methods are generally more desirable. A typical SEC system can quickly (in about half an hour) give polymer chemists information on the size and polydispersity of the sample.

Ion exchange Chromatography

Ion-exchange chromatography (or ion chromatography) is a process that allows the separation of ions and polar molecules based on the charge properties of the molecules. It can be used for almost any kind of charged molecule including large proteins, small nucleotides and amino acids, with the experimental solution to be separated collectively known as the analyte. It is often used as a first step in protein purification. Ion exchange chromatography retains analyte molecules based on coulombic (ionic) interactions. The stationary phase surface displays ionic functional groups that interact with analyte ions of opposite charge.

This type of chromatography is further subdivided into cation exchange chromatography and anion exchange chromatography:

1. Cation exchange chromatography retains positively charged cations because the stationary phase displays a negatively charged functional group such as a phosphoric acid.

2. Anion exchange chromatography retains negatively charged anions using positively charged functional group such as a quaternary ammonium cation.

Proteins have numerous functional groups that can have both positive and negative charges. Ion exchange chromatography separates proteins according to their net charge, which is dependent on the composition of the mobile phase. By adjusting the pH or the ionic concentration of the mobile phase, various protein molecules can be separated.

For example, if a protein has a net positive charge at pH 7, then it will bind to a column of negatively-charged beads, whereas a negatively charged protein would not. By changing the pH so that the net charge on the protein is negative, it too will be eluted.

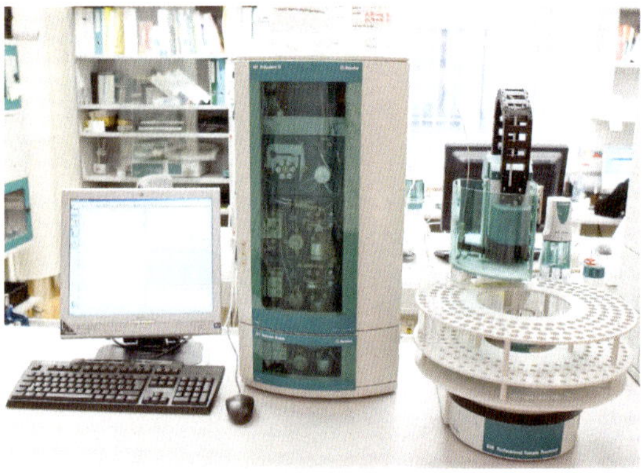

Ion exchange chromatography set-up

Elution by changing the ionic strength of the mobile phase is a more subtle effect — it works as ion from the mobile phase and will interact with the immobilized ion in preference over those on the stationary phase. This "shields" the stationary phase from the protein (and vice versa) and allows the protein to elute.

A sample is introduced, either manually or with an auto sampler, into a sample loop of known volume. A buffered aqueous solution known as the mobile phase carries the sample from the loop onto a column that contains some form of stationary phase material. This is typically a resin or gel matrix consisting of agarose or cellulose beads with covalently bonded charged functional groups.

The target analytes (anions or cations) are retained on the stationary phase but can be eluted by increasing the concentration of a similarly charged species that will displace the analyte ions from the stationary phase. For example, in cation exchange chromatography, the positively charged analyte could be displaced by the addition of positively charged sodium ions.

The analytes of interest must then be detected by some means, typically by conductivity or UV/Visible light absorbance. In order to control an IC system, a Chromatography Data System (CDS) is usually needed. In addition to IC systems, some of these CDSs can also control Gas Chromatography (GC) and HPLC systems.

Bio-affinity Chromatography

This chromatographic process relies on the property of biologically active substances to form stable, specific, and reversible complexes. The formation of these complexes involves the participation of common molecular forces such as the van der Waals interaction, electrostatic interaction, dipole- dipole interaction, hydrophobic interaction, and the hydrogen bond. An efficient, bio-specific bond is formed by a simultaneous and concerted action of several of these forces in the complementary binding sites.

Parameters

There are different parameters upon which the separation by HPLC relies. Some of the important parameters are discussed below:

Internal Diameter

The internal diameter (ID) of an HPLC column is a critical aspect that determines quantity of analyte that can be loaded onto the column and also influences sensitivity. Larger columns are usually seen in industrial applications such as the purification of a drug product for later use. Low ID columns have improved sensitivity and lower solvent consumption at the expense of loading capacity.

1. Larger ID columns (over 10 mm) are used to purify usable amounts of material because of their large loading capacity.
2. Analytical scale columns (4.6 mm) have been the most common type of columns, though smaller columns are rapidly gaining popularity. They are used in traditional quantitative analysis of samples and often use a UV-Vis absorbance detector.
3. Narrow-bore columns (1-2 mm) are used for applications when more sensitivity is desired either with special UV-Vis detectors, fluorescence detection or with other detection methods like liquid chromatography-mass spectrometry
4. Capillary columns (under 0.3 mm) which are used almost exclusively with alternative detection means such as mass spectrometry. They are usually made from fused silica capillaries, rather than the stainless steel tubing that larger columns employ.

Particle Size

Most traditional HPLC is performed with the stationary phase attached to the outside of small spherical silica particles (very small beads). These particles come in a variety of sizes with 5 μm beads being the most common. Smaller particles generally provide more surface area and better separations, but the pressure required for optimum linear velocity increases by the inverse of the particle diameter squared.

This means that changing to particles that are half as big in the same size of column will double the performance, but increase the required pressure by a factor of four. Larger particles are more often used in non-HPLC applications such as solid-phase extraction.

Pore Size

Many stationary phases are porous to provide greater surface area. Small pores provide greater surface area while larger pore size has better kinetics especially for larger analytes. For example, a protein which is only slightly smaller than a pore might enter the pore but not easily leave once inside.

Pump Pressure

Pumps vary in pressure capacity, but their performance is measured on their ability to yield a consistent and reproducible flow rate. Pressure may reach as high as 6000 lbf/in^2 (~40 MPa, or about 400 atmospheres). Modern HPLC systems have been improved to work at much higher pressures, and therefore, be able to use much smaller particle sizes in the columns (< 2 micrometres).

These "Ultra High Performance Liquid Chromatography" systems or UHPLCs can work at up to 15,000 lbf/in^2 (-100 MPa or about 1000 atmospheres). Note that the term "UPLC", sometimes found instead is a trademark of Waters Corporation and not the name for the technique in general.

Fast Protein Liquid Chromatography

Fast Protein Liquid Chromatography, usually referred to as FPLC, is a form of column chromatography used to separate or purify proteins from complex mixtures. It is very commonly used in biochemistry and enzymology. Columns used with an FPLC can separate macromolecules based on size, charge distribution, hydrophobicity, or bio-recognition (as with affinity chromatography).

Typical columns used for protein purification include:

1. Ion exchange chromatography, cation or anion — separates proteins based on surface-charges
2. Gel filtration chromatography, also known as Gel permeation chromatography or desalting — separates proteins based on size.
3. Reversed phase or hydrophobic interaction — separates based on hydrophobicity which depends heavily on surface area.
4. Affinity chromatography — columns which purify based on ligand affinity, such as a His-tagged protein uses a nickel column.

Supercritical Fluid Chromatography (SFC)

It is a robust and easy-to-use form of normal phase chromatography ideally suited to the analysis and purification of low to moderate molecular weight, thermally labile molecules. It is especially suited to the separation of chiral compounds. Similar to high performance liquid chromatography (HPLC), SFC typically utilizes carbon dioxide as the mobile phase; therefore, the entire chromatographic flow path must be pressurized.

In addition, SFC metering pumps require that the pump head be kept cold in order to maintain the carbon dioxide in a supercritical state, where it can be effectively metered at some specified flow rate. The chemist sets mobile phase flow rate, composition, and column temperature. In addition, SFC provides an additional control parameter, pressure, which the chemist similarly sets through the keyboard. From an operational standpoint, SFC is as simple and robust as HPLC.

Any molecule that will dissolve in methanol or a less polar solvent is an ideal candidate for SFC. SFC can even separate polar solutes. Many strong bases that are difficult to separate by other techniques separate rapidly and efficiently with good peak shapes.

Affinity Chromatography

It is a chromatographic method of separating biochemical mixtures, based on a highly specific biologic interaction such as that between antigen and antibody, enzyme and substrate, or receptor and ligand. Affinity chromatography combines the size fractionation capability of gel permeation chromatography with the ability to design a stationary phase that reversibly binds to a known subset of molecules.

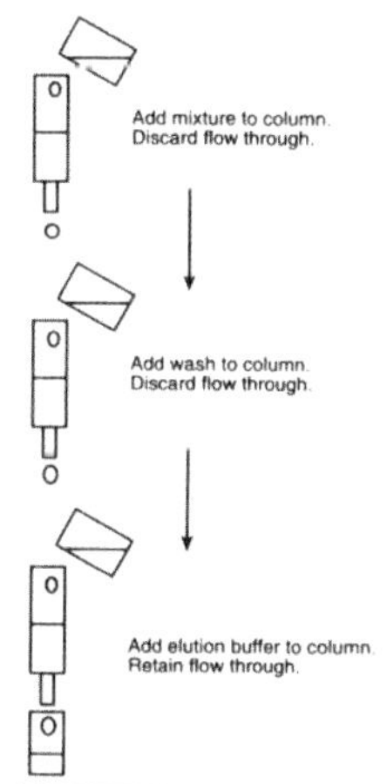

Fig. 9.12: Column chromatography

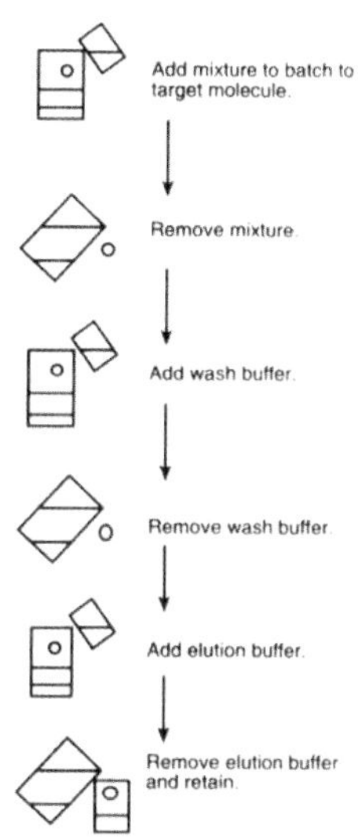

Affinity chromatography can be used to:

1. Purify and concentrate a molecule from a mixture into a buffering solution.
2. Reduce the amount of a molecule in a mixture.
3. Discern what biological compounds bind to a particular molecule, such as drugs.

Usually the starting point is an undefined heterogeneous group of molecules in solution, such as a cell lysate, growth medium or blood serum. The molecule of interest will have a well-known and defined property which can be exploited during the affinity purification process. The process itself can be thought of as an entrapment, with the target molecule becoming trapped on a solid or stationary phase or medium.

The other molecules in solution will not become trapped as they do not possess this property. The solid medium can then be removed from the mixture, washed and the target molecule released from the entrapment in a process known as elution.

Binding to the solid phase may be achieved by column chromatography, whereby the solid medium is packed onto a chromatography column, the initial mixture run through the column to allow binding, a wash buffer run through the column and the elution buffer subsequently applied to the column and collected. These steps are usually done at ambient pressure (as opposed to HPLC or FPLC).

Alternatively binding may be achieved using a batch treatment, by adding the initial mixture to the solid phase in a vessel, mixing, separating the solid phase (by centrifugation, for example), removing the liquid phase, washing, re-centrifuging, adding the elution buffer, re-centrifuging and removing the eluate. Sometimes a hybrid method is employed, the binding is done by the batch method, then the solid phase with the target molecule bound is packed onto a column and washing and elution are done on the column.

A third method, expanded bed adsorption, which combines the advantages of the two methods mentioned above, has also been developed. The solid phase particles are placed in a column where liquid phase is pumped in from the bottom and exits at the top. The gravity of the particles ensures that the solid phase does not exit the column with the liquid phase.

Application

Affinity chromatography can be used in a number of applications, including nucleic acid purification, protein purification from cell free extracts and antibody purification from blood serum.

Antibody Affinity

Another use of the procedure is the affinity purification of antibodies from blood serum. If serum is known to contain antibodies against a specific antigen (for example, if the serum comes from an organism immunized against the antigen concerned), then it can be used for the affinity purification of that antigen. For example, if an organism is immunized against a GST-fusion protein it will produce antibodies against the fusion-protein, and possibly antibodies against the GST tag as well. The protein can then be covalently coupled to a solid support such as agarose.

For thoroughness the GST protein and the GST-fusion protein can each be coupled separately. The serum is initially allowed to bind to the GST affinity matrix. This will remove antibodies against the GST part of the fusion protein. The serum is then separated from the solid support and allowed to bind to the GST-fusion protein matrix. This allows any antibodies that recognize the antigen to be captured on the solid support.

Elution of the antibodies of interest is most often achieved using a low pH buffer such as glycine, pH 2.8. The eluate is collected into a neutral tris or phosphate buffer to neutralize the low pH elution buffer and halts any degradation of the antibody's activity. This is a nice example as affinity purification is used to purify the initial GST-fusion protein to remove the undesirable anti-GST antibodies from the serum and to purify the target antibody.

Immobilized Metal Ion Affinity Chromatography

Immobilized metal ion affinity chromatography (IMAC) is based on the specific coordinate covalent binding of amino acids, allowing proteins with an affinity for metal ions to be retained in a column containing immobilized metal ions, such as cobalt, nickel, copper, zinc, or iron ions. Many naturally occurring proteins do not have an affinity for metal ions; therefore, recombinant DNA techniques are used to introduce this property into a protein of interest.

Methods used to elute the protein of interest include changing the pH, or adding a specific molecule, such as imidazole. Similarly, the plant protein concanavalin A is able to bind with glucose immobilized on a column matrix. The concanavalin molecule is separated from the glucose by increasing the glucose concentration, displacing the concanavalin from the immobilized glucose.

Recombinant Proteins

Possibly the most common use of affinity chromatography is for the purification of recombinant proteins. Proteins with a known affinity are tagged in order to aid their purification. The protein may have been genetically modified so as to allow it to be selected for affinity binding; this is known as a fusion protein. Tags include His-tags and GST (glutathione-S-transferase) tags.

His_6-tags have an affinity for nickel or cobalt ions which are coordinated with NTA for the purposes of solid medium entrapment. For elution, an excess amount of a compound able to act as a nickel ligand, such as imidazole, is used. GST has an affinity for glutathione – commercially available immobilized as glutathione agarose. For elution, excess glutathione is used to displace the tagged protein.

Reversed-Phase Chromatography

Reverse-phase chromatography (RPC) includes any chromatographic method that uses a non-polar stationary phase. All of the mathematical and experimental considerations used in other chromatographic methods apply (i.e., separation resolution proportional to the column length and inversely proportional to the column width).

Reversed-phase column chromatography is widely used in the pharmaceutical, chemical, and biochemical industry for separating molecules of small molecular weight. In more recent years RPC has been used to separate larger molecules. Any inert non-polar substance that achieves sufficient packing can be used for reversed-phase chromatography. Common examples include larger hydrocarbons, diphenyl, and divinylbenzene.

Typically mobile phases are made of water or a polar organic compound such as acetonitrile or the lighter alcohols. Buffer solutions are used to control the protonation of functional groups. For example, a buffer of high pH will encourage elution of alcohols because they are charged in basic solutions and, therefore, prefer to be in the mobile phase. For the same reason, very acidic solutions encourage the elution of nitrogen containing molecules. Using multiple buffers allows for selective elution of a wide variety of chemicals.

Two-Dimensional Chromatography

By using an additional physicochemical (Chemical classification) criterion for separation of the mixture of analytes (sample), the resolution and quality of chromatographic separation can be increased. As a result, higher specificity concerning the separational capability of the chromatographic technique is obtained, allowing separation and preparation or analysis of compounds indistinguishable by one-dimensional chromatography.

In Gas-Phase Chromatography, two-dimensional separation is achieved by coupling a second, short column to the first long column. Coupling is achieved by different techniques, for example, shock-freezing the elutes in order of elution from the first column at fixed time-intervals, and then reheating them in order of elution, releasing them into the second column. The time of traversal through the second column needs to be shorter than the time remaining until the next sample is reheated to prevent compound build-up and to fully exploit the separational capability.

Pyrolysis Gas Chromatography

Pyrolysis-chromatography is a potent analytical tool able to thermally crack (fragment) essentially non-volatile molecules into fragments suitable for chromatographic analysis. The technique enables a reproducible and characteristic "fingerprint" to be generated of a non-volatile sample. The technique can be applied to such varied tasks as bacterial strain differentiation and forensic characterisation of paints, polymers and fibre cross-matching.

Type of Chromatography	Applications in the Real World	Why and What is it
Liquid Chromatography	testing water samples to look for pollution	Used to analyze metal ions and organic compounds in solutions. It uses liquids which may incorporate hydrophilic, insoluble molecules.
Gas Chromatography	Detecting bombs in airports, identifying and quantifying such drugs as alcohol, being used in forensics to compare fibres found on a victim	Used to analyze volatile gases. Helium is used to move the gaseous mixture through a column of absorbent material.
Thin-Layer Chromatography	Detecting pesticide or insecticide residues in food, also used in forensics to analyze the dye composition of fibres	Uses an absorbent material on flat glass plates. This is a simple and rapid method to check the purity of the organic nd.
Paper Chromatography	separating amino acids and an ions, RNA fingerprinting, separating and testing histamines, antibiotics	The most common type of chromatography. The paper is the stationary phase. This uses capillary action to pull the solutes up through the paper and separate the solutes.

Adsorption Chromatography

Adsorption chromatography is probably one of the oldest types of chromatography around. It utilizes a mobile liquid or gaseous phase that is adsorbed onto the surface of a stationary solid phase. The equilibration between the mobile and stationary phases accounts for the separation of different solutes.

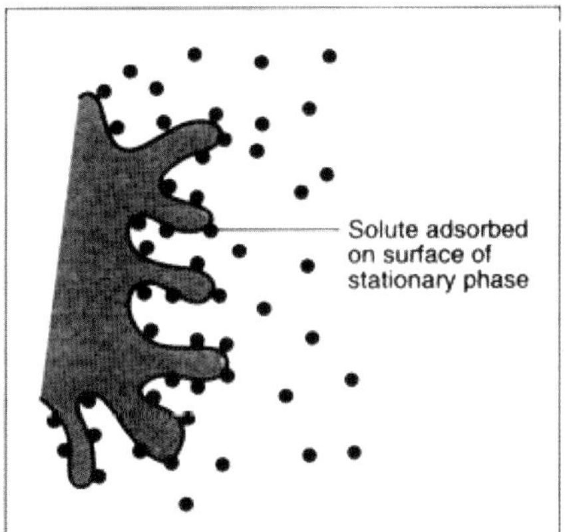

Adsorption chromatography

Partition Chromatography

This form of chromatography is based on a thin film formed on the surface of a solid support by a liquid stationary phase. Solute equilibrates between the mobile phase and the stationary liquid.

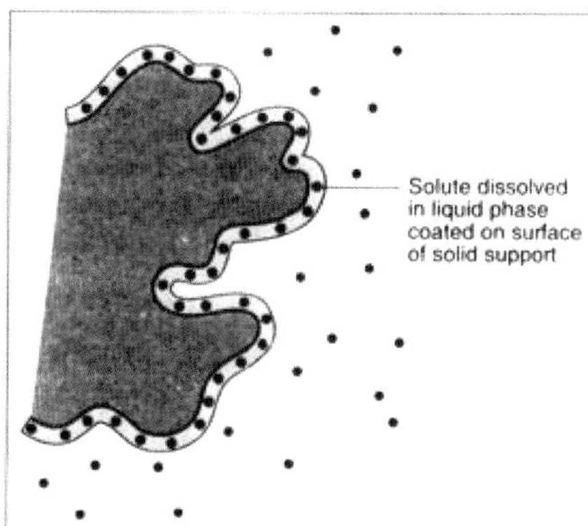

Ion Exchange Chromatography

In this type of chromatography, the use of a resin (the stationary solid phase) is used to covalently attach anions or cations onto it. Solute ions of the opposite charge in the mobile liquid phase are attracted to the resin by electrostatic forces.

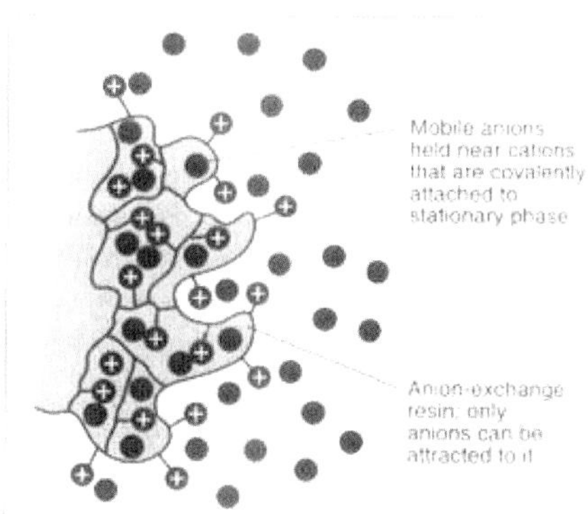

Ion-exchange chromatography

Molecular Exclusion Chromatography

Also known as gel permeation or gel filtration, this type of chromatography lacks an attractive interaction between the stationary phase and solute. The liquid or gaseous phase passes through a porous gel which separates the molecules according to its size. The pores are normally small and exclude the larger solute molecules, but allow smaller molecules to enter the gel, causing them to flow through a larger volume. This causes the larger molecules to pass through the column at a faster rate than the smaller ones.

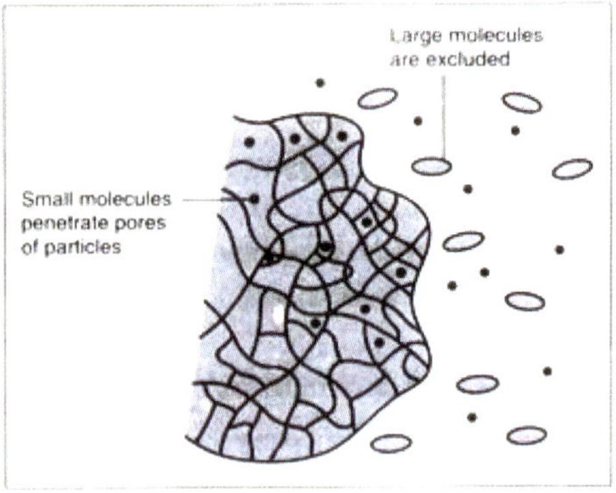

Affinity Chromatography

This is the most selective type of chromatography employed. It utilizes the specific interaction between one kind of solute molecule and a second molecule that is immobilized on a stationary phase. For example, the immobilized molecule may be an antibody to some specific protein. When solute containing a mixture of proteins are passed by this molecule, only the specific protein is reacted to this antibody, binding it to the stationary phase. This protein is later extracted by changing the ionic strength or pH.

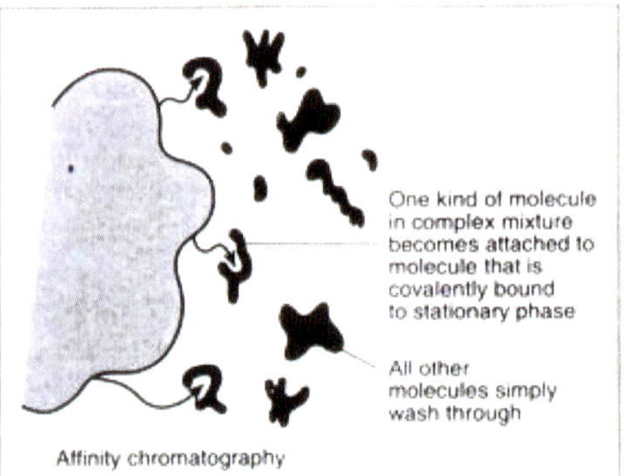

Affinity chromatography

Counter-Current Chromatography

CCC or Partition Chromatography is a category of liquid-liquid chromatography techniques. Chromatography in general is used to separate components of a mixture based on their differing affinities for mobile and stationary phases of a column. The components can then be analyzed separately by various sorts of detectors which may or may not be integrated into an apparatus. In liquid-liquid chromatography, both the mobile and stationary phases are liquid.

In contrast, standard column chromatography uses a solid stationary phase and a liquid mobile phase, while gas chromatography uses a liquid stationary phase on a solid support and a gaseous mobile phase. By eliminating solid supports, permanent adsorption of the analyte onto the column is avoided, and a near 100% recovery of the analyte can be achieved.

The instrument is also easily switched between various modes of operation simply by changing solvents. With liquid-liquid chromatography, researchers are not limited by the composition of the columns commercially available for their instrument. Nearly any pair of immiscible solutions can be used in liquid-liquid chromatography, and most instruments can be operated in standard or reverse-phase modes. Solvent costs are also generally cheaper than that HPLC, and the cost of purchasing and disposing of solid adsorbents is completely eliminated.

Another advantage is that experiments conducted in the lab can easily be scaled to industrial volumes. When GC or HPLC is done with large volumes, resolution is lost due to issues with surface-to-volume ratios and flow dynamics; this is avoided when both phases are liquid.

CCC can be thought of as occurring in three stages: mixing, settling, and separation (although they often occur continuously). Mixing of the phases is necessary so that the interface between them has a large area, and the analyte can move between the phases according to its partition coefficient.

A partition coefficient is a ratio of the amount of analyte found in each of the solvents at equilibrium and is related to the analyses' affinity for one over the other. The mobile phase is mixing with them settling from the stationary phase throughout the column.

The degree of stationary phase retention (inversely proportional to the amount of stationary phase loss or "bleed" in the course of a separation) is a crucial parameter. Higher quality instruments have greater stationary phase retention. The settling time is a property of the solvent system and the sample matrix, both of which greatly influence stationary phase retention.

There are certain modifications of CCC. Major of them are:

1. Droplet Counter-current Chromatography (DCCC)

Droplet CCC is the oldest form of CCC. It uses only gravity to move the mobile phase through the stationary phase. In descending mode, droplets of the more dense mobile phase and sample are allowed to fall through a column of the lighter stationary phase using only gravity. If a less dense mobile phase is used it will rise through the stationary phase, this is called ascending mode.

The eluent from one column is transferred to another; the more columns that are used, the more theoretical plates can be achieved. The disadvantage of DCCC is that flow rates are low, and poor mixing is achieved for most binary solvent systems, which make this technique both time-consuming and inefficient.

2. Centrifugal Partition Chromatography (CPC)

CPC uses centrifugal force to speed separation and achieve greater flow rates than DCCC (which relies on gravity) (e.g., the Kromaton FCPC). The columns are cut into a rotor, oriented out from the middle, and connected by channels. The rotor is filled with the stationary phase, and the mobile phase is pumped through it as the rotor spins.

Early CPC instruments had the disadvantage of requiring a rotary seal that frequently needed to be replaced, but a new generation of instruments is now reliable. CPC can also be operated in either descending or ascending mode, where the direction is relative to the force generated by the rotor rather than gravity.

3. High-Speed Counter-current Chromatography

The modern era of CCC began with the development by Dr. Yoichiro Ito of the planetary centrifuge and the many possible column geometries it can support. These clever devices make use of a little-known means of making non-rotating connections between the stator and the rotor of a centrifuge. (It is beyond the scope of this discussion to describe the method of accomplishing this. Any of the several books available on CCC discuss it thoroughly.)

Functionally, the high-speed CCC consists of a helical coil of inert tubing which rotates on its planetary axis and simultaneously rotates eccentrically about another solar axis. (These axes can be made to coincide, but the most common or type J CCC is discussed here.) The effect is to create zones of mixing and zones of settling which progress along the helical coil at dizzying speed. This produces a highly favorable environment for chromatography.

There are numerous potential variants upon this instrument design. The most significant of these is the toroidal CCC (e.g., the Pharma-Tech TCC-1000). This instrument does not employ planetary motion. In some respects it is very alike CPC, but retains the advantage of not needing rotary seals.

It also employs a capillary tube instead of the larger-diameter tubes employed in the helices of the other CCC models. This capillary passage makes the mixing of two phases very thorough, despite the lack of shaking or other mixing forces. This instrument provides rapid analytical scale separations, which can nonetheless be scaled up to either of the larger-scale CCC instruments.

Modes of Operation

CCC can be operated in different modes depending on the type of separation required and sample to be separated.

These different operation modes of CCC are as under:

i. Head to Tail

The denser phase is pumped through as the mobile phase. Derived from terminology for Archimedean screw force.

ii. Tail to Head

The less dense phase is used as the mobile phase.

iii. Dual Mode

The mobile and stationary phases are reversed part way through the run.

iv. Gradient Mode

The concentration of one or more components in the mobile phase is varied throughout the run to achieve optimal resolution across a wider range of polarities. For example, a methanol-water gradient may be employed using pure heptane as the stationary phase. This is not possible with all binary systems due to excessive loss of stationary phase.

v. Elution Extrusion Mode (EECCC)

The mobile phase is extruded after a certain point by switching the phase being pumped into the system. For example, during the elution portion of a separation using an EtOAc-water system running head to tail, the aqueous mobile phase is being pumped into the system. In order to switch to extrusion mode, organic phase is pumped into the system.

This can be accomplished either with a valve on the inlet of single pump, or ideally with an integrated system of two or three pumps, each dedicated either to a single phase of a binary mixture, or to an intermediate wash solvent. This also allows for good resolution of compounds with high mobile-phase affinities. It requires only one column volume of solvent and leaves the column full of fresh stationary phase.

vi. pH Zone Refining

Acidic and basic solvents are used to elute analytes based on their pKa.

Example of Metal Extraction

Aluminium Extraction

Aluminium is too high in the electrochemical series (reactivity series) to extract it from its ore using carbon reduction. The temperatures needed are too high to be economic.

Instead, it is extracted by electrolysis. The ore is first converted into pure aluminium oxide by the Bayer Process, and this is then electrolysed in solution in molten cryolite - another aluminium compound. The aluminium oxide has too high a melting point to electrolyse on its own.

Aluminium Ore

The usual aluminium ore is bauxite. Bauxite is essentially an impure aluminium oxide. The major impurities include iron oxides, silicon dioxide and titanium dioxide.

Purifiying the Aluminium Oxide - the Bayer Process

Reaction with Sodium Hydroxide Solution

Crushed bauxite is treated with moderately concentrated sodium hydroxide solution. The concentration, temperature and pressure used depend on the source of the bauxite and exactly what form of aluminium oxide it contains. Temperatures are typically from 140 °C to 240 °C; pressures can be up to about 35 atmospheres.

High pressures are necessary to keep the water in the sodium hydroxide solution liquid at temperatures above 100 °C. The higher the temperature, the higher the pressure needed.

With hot concentrated sodium hydroxide solution, aluminium oxide reacts to give a solution of sodium tetrahydroxoaluminate.

$$Al_2O_3 + 2\,NaOH + 3\,H_2O \rightarrow 2\,NaAl(OH)_4$$

The impurities in the bauxite remain as solids. For example, the other metal oxides present tend not to react with the sodium hydroxide solution and so remain unchanged. Some of the silicon dioxide reacts, but goes on to form a sodium aluminosilicate which precipitates out.

All of these solids are separated from the sodium tetrahydroxoaluminate solution by filtration. They form a "red mud" which is just stored in huge lagoons.

Precipitation of Aluminium Hydroxide

The sodium tetrahydroxoaluminate solution is cooled, and "seeded" with some previously produced aluminium hydroxide. This provides something for the new aluminium hydroxide to precipitate around.

$$NaAl(OH)_4 \rightarrow Al(OH)_3 + NaOH$$

Formation of Pure Aluminium Oxide

Aluminium oxide (sometimes known as alumina) is made by heating the aluminium hydroxide to a temperature of about 1100 – 1200 °C.

$$2\,Al(OH)_3 \rightarrow Al_2O_3 + 3\,H_2O$$

Conversion of the Aluminium Oxide into Aluminium by Electrolysis

The aluminium oxide is electrolysed in solution in molten cryolite, Na_3AlF_6. Cryolite is another aluminium ore, but is rare and expensive, and most is now made chemically.

Electrolysis Cell

The diagram shows a very simplified version of an electrolysis cell.

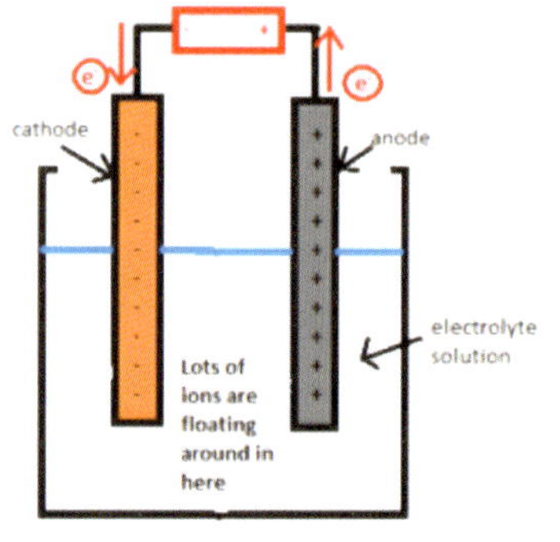

The electrolysis cell

Although the carbon lining of the cell is labelled as the cathode, the effective cathode is mainly the molten aluminium that forms on the bottom of the cell.

Molten aluminium is syphoned out of the cell from time to time, and new aluminium oxide added at the top.

The cell operates at a low voltage of about 5 - 6 volts, but at huge currents of 100,000 amps or more. The heating effect of these large currents keeps the cell at a temperature of about 1000 °C.

The Electrode Reactions

These are very complicated - in fact one source I've looked at says that they aren't fully understood. For chemistry purposes at this level, they are always simplified.

This is the simplification:

Aluminium is released at the cathode. Aluminium ions are reduced by gaining 3 electrons.

$$Al^{3}+3e^{-} \rightarrow Al$$

Oxygen is produced initially at the anode.

$$2O^{2-} \rightarrow O^{2}+4e^{-}$$

However, at the temperature of the cell, the carbon anodes burn in this oxygen to give carbon dioxide and carbon monoxide.

Continual replacement of the anodes is a major expense.

Some Economic and Environmental Considerations

This topic is designed to give you a brief idea of the sort of economic and environmental issues involved with the extraction of aluminium.

Economic Considerations

- The high cost of the process because of the huge amounts of electricity it uses. This is so high because to produce 1 mole of aluminium which only weighs 27 g you need 3 moles of electrons. You are having to add a lot of electrons (because of the high charge on the ion) to produce a small mass of aluminium (because of its low relative atomic mass).
- Energy and material costs in constantly replacing the anodes.
- Energy and material costs in producing the cryolite, some of which gets lost during the electrolysis.

Environmental Problems in Mining and Transporting the Bauxite

- Loss of landscape due to mining, processing and transporting the bauxite.
- Noise and air pollution (greenhouse effect, acid rain) involved in these operations.

Extracting Aluminium from the Bauxite

- Loss of landscape due to the size of the chemical plant needed, and in the production and transport of the electricity.
- Noise.
- Atmospheric pollution from the various stages of extraction. For example: carbon dioxide from the burning of the anodes (greenhouse effect); carbon monoxide (poisonous); fluorine (and fluorine compounds) lost from the cryolite during the electrolysis process (poisonous).
- Pollution caused by power generation (varying depending on how the electricity is generated.)
- Disposal of red mud into unsightly lagoons.
- Transport of the finished aluminium.

Recycling

- Saving of raw materials and particularly electrical energy by not having to extract the aluminium from the bauxite. Recycling aluminium uses only about 5% of the energy used to extract it from bauxite.
- Avoiding the environmental problems in the extraction of aluminium from the bauxite.
- Not having to find space to dump the unwanted aluminium if it wasn't recycled.
- (Offsetting these to a minor extent) Energy and pollution costs in collecting and transporting the recycled aluminium.

Uses of Aluminium

Aluminium is usually alloyed with other elements such as silicon, copper or magnesium. Pure aluminium isn't very strong, and alloying it adds to it strength.

Aluminium is especially useful because it;

- Has a low density;
- Is strong when alloyed;
- Is a good conductor of electricity;
- Has a good appearance;
- Resists corrosion because of the strong thin layer of aluminium oxide on its surface. This layer can be strengthened further by anodising the aluminium.

  Anodising essentially involves etching the aluminium with sodium hydroxide solution to remove the existing oxide layer, and then making the aluminium article the anode in an electrolysis of dilute sulphuric acid. The oxygen given of at the anode reacts with the aluminium surface, to build up a film of oxide up to about 0.02 mm thick.

As well as increasing the corrosion resistance of the aluminium, this film is porous at this stage and will also take up dyes. (It is further treated to make it completely non-porous afterwards.) That means that you can make aluminium articles with the colour built into the surface.

Some uses include:

Aluminium is used for	Because
aircraft	light, strong, resists corrosion
other transport such as ships' superstructures, container vehicle bodies, tube trains (metro trains)	light, strong, resists corrosion
overhead power cables (with a steel core to strengthen them)	light, resists corrosion, good conductor of electricity
saucepans	light, resists corrosion, good appearance, good conductor of heat

Iron Extraction

The common ores of iron are both iron oxides, and these can be reduced to iron by heating them with carbon in the form of coke. Coke is produced by heating coal in the absence of air. Coke is cheap and provides both the reducing agent for the reaction and also the heat source. The most commonly used iron ores are hematite, Fe_2O_3, and magnetite, Fe_3O_4.

Blast Furnace

The significant reactions occuring within the Blast Furnace can be described via the following steps showing how the reducing agent varies depending on the height in the furnace (i.e. on the Temperature).

At 500 °C

$$3Fe_2O_3 + CO \rightarrow 2Fe_3O_4 + CO_2$$
$$Fe_2O_3 + CO \rightarrow 2FeO + CO_2$$

At 850 °C

$$Fe_3O_4 + CO \rightarrow 3FeO + CO_2$$

At 1000 °C

$$FeO + CO \rightarrow Fe + CO_2$$

At 1300 °C

$$CO_2 + C \rightarrow 2CO$$

At 1900 °C

$$C + O_2 \rightarrow CO_2$$
$$FeO + C \rightarrow Fe + CO$$

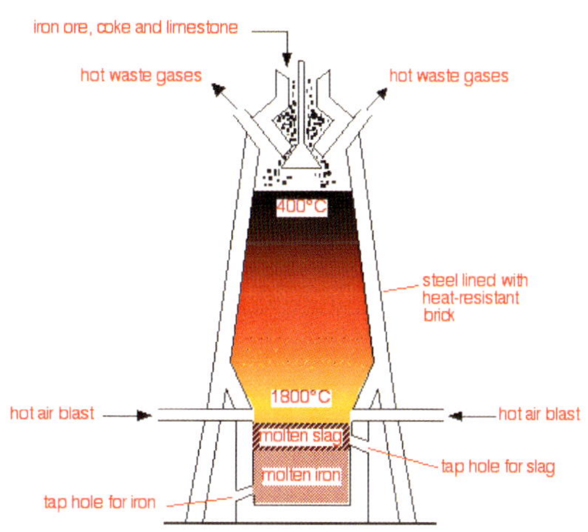

The air blown into the bottom of the furnace is heated using the hot waste gases from the top. Heat energy is valuable, and it is important not to waste any. The coke (essentially impure carbon)

burns in the blast of hot air to form carbon dioxide - a strongly exothermic reaction. This reaction is the main source of heat in the furnace.

$$C+O_2 \rightarrow 2CO_2$$

It is the carbon monoxide which is the main reducing agent in the furnace.

$$Fe_2O_3+3CO \rightarrow 2Fe+3CO_2$$

In the hotter parts of the furnace, the carbon itself also acts as a reducing agent. Notice that at these temperatures, the other product of the reaction is carbon monoxide, not carbon dioxide.

$$Fe_2O_3+3C \rightarrow 2Fe+3CO$$

The temperature of the furnace is hot enough to melt the iron which trickles down to the bottom where it can be tapped off.

Function of the Limestone

Iron ore is not pure iron oxide - it also contains an assortment of rocky material that would not melt at the temperature of the furnace, and would eventually clog it up. The limestone is added to convert this into slag which melts and runs to the bottom. The heat of the furnace decomposes the limestone to give calcium oxide.

$$CaCO_3 \rightarrow CaO+CO_2$$

This is an endothermic reaction, absorbing heat from the furnace. It is therefore important not to add too much limestone because it would otherwise cool the furnace. Calcium oxide is a basic oxide and reacts with acidic oxides such as silicon dioxide present in the rock. Calcium oxide reacts with silicon dioxide to give calcium silicate.

$$CaO+SiO_2 \rightarrow CaSiO_3$$

The calcium silicate melts and runs down through the furnace to form a layer on top of the molten iron. It can be tapped off from time to time as slag. Slag is used in road making and as "slag cement" - a final ground slag which can be used in cement, often mixed with Portland cement.

Cast Iron

The molten iron from the bottom of the furnace can be used as cast iron. Cast iron is very runny when it is molten and doesn't shrink much when it solidifies. It is therefore ideal for making castings - hence its name. However, it is very impure, containing about 4% of carbon. This carbon makes it very hard, but also very brittle. If you hit it hard, it tends to shatter rather than bend or dent. Cast iron is used for things like manhole covers, guttering and drainpipes, cylinder blocks in car engines, Aga-type cookers, and very expensive and very heavy cookware.

Steel

Most of the molten iron from a Blast Furnace is used to make one of a number of types of steel. There isn't just one substance called steel - they are a family of alloys of iron with carbon or various metals.

Steel-making: Basic Oxygen Process

Impurities in the iron from the Blast Furnace include carbon, sulfur, phosphorus and silicon. These have to be removed.

- Removal of sulfur: Sulfur has to be removed first in a separate process. Magnesium powder is blown through the molten iron and the sulphur reacts with it to form magnesium sulfide. This forms a slag on top of the iron and can be removed.

  $Mg + S \rightarrow MgS$

- Removal of carbon: The still impure molten iron is mixed with scrap iron (from recycling) and oxygen is blown on to the mixture. The oxygen reacts with the remaining impurities to form various oxides. The carbon forms carbon monoxide. Since this is a gas it removes itself from the iron. This carbon monoxide can be cleaned and used as a fuel gas.

- Removal of other elements: Elements like phosphorus and silicon react with the oxygen to form acidic oxides. These are removed using quicklime (calcium oxide) which is added to the furnace during the oxygen blow. They react to form compounds such as calcium silicate or calcium phosphate which form a slag on top of the iron.

Types of Iron and Steel

Cast iron has already been mentioned above. This topic deals with the types of iron and steel which are produced as a result of the steel-making process.

- Wrought iron: If all the carbon is removed from the iron to give high purity iron, it is known as wrought iron. Wrought iron is quite soft and easily worked and has little structural strength. It was once used to make decorative gates and railings, but these days mild steel is normally used instead.

- Mild steel: Mild steel is iron containing up to about 0.25% of carbon. The presence of the carbon makes the steel stronger and harder than pure iron. The higher the percentage of carbon, the harder the steel becomes. Mild steel is used for lots of things - nails, wire, car bodies, ship building, girders and bridges amongst others.

- High carbon steel: High carbon steel contains up to about 1.5% of carbon. The presence of the extra carbon makes it very hard, but it also makes it more brittle. High carbon steel is used for cutting tools and masonry nails (nails designed to be driven into concrete blocks or brickwork without bending). You have to be careful with high carbon steel because it tends to fracture rather than bend if you mistreat it.

- Special steels: These are iron alloyed with other metals. For example:

	iron mixed with	special properties	uses include
stainless steel	chromium and nickel	resists corrosion	cutlery, cooking utensils, kitchen sinks, industrial equipment for food and drink processing
titanium steel	titanium	withstands high temperatures	gas turbines, spacecraft
manganese steel	manganese	very hard	rock-breaking machinery, some railway track (e.g. points), military helmets

Environmental Considerations

Environmental Problems in Mining and Transporting the Raw Materials

- Loss of landscape due to mining, processing and transporting the iron ore, coke and limestone.
- Noise and air pollution (greenhouse effect, acid rain) involved in these operations.

Extracting Iron from the Ore

- Loss of landscape due to the size of the chemical plant needed.
- Noise.
- Atmospheric pollution from the various stages of extraction. For example: carbon dioxide (greenhouse effect); carbon monoxide (poisonous); sulfur dioxide from the sulfur content of the ores (poisonous, acid rain).
- Disposal of slag, some of which is just dumped.
- Transport of the finished iron.

Recycling

- Saving of raw materials and energy by not having to first extract the iron from the ore.
- Avoiding the pollution problems in the extraction of iron from the ore.
- Not having to find space to dump the unwanted iron if it wasn't recycled.
- (Offsetting these to a minor extent) Energy and pollution costs in collecting and transporting the recycled iron to the steel works.

Copper Extraction

Important Ores of Copper

- Copper pyrite or chalcopyrite $(CuFeS_2)$.
- Chalocite (Cu_2S) or copper glance.
- Malachite green $[CuCO_3.Cu(OH)_2]$.
- Azurite blue $[2CuCO_3.Cu(OH)_2]$.
- Bornite $(3Cu_2S.Fe2S_3)$ or peacock ore.
- Melaconite (CuO) etc.

Extraction of Copper from Sulphide Ore

Large amount of copper are obtained from copper pyrite (CuFeS2) by smelting. Ores containing 4% or more copper are treated by smelting process. Very poor ores are treated by hydro-metallurgical process.

Extraction of Copper by Smelting Process

Following steps are involved in the extraction of copper.

Crashing and Grinding

The ore is crushed then ground into powder.

Concentration

The finely crushed ore is concentrated by Froth-Floatation process. The finely crushed ore is suspended in water containing a little amount of pine oil. A blast of air is passed through the suspension. The particles get wetted by the oil and float as a froth which is skimmed. The gangue (unwanted materials) sinks to the bottom then removed. Productions of 25% approximate Cu containing powder in this stage.

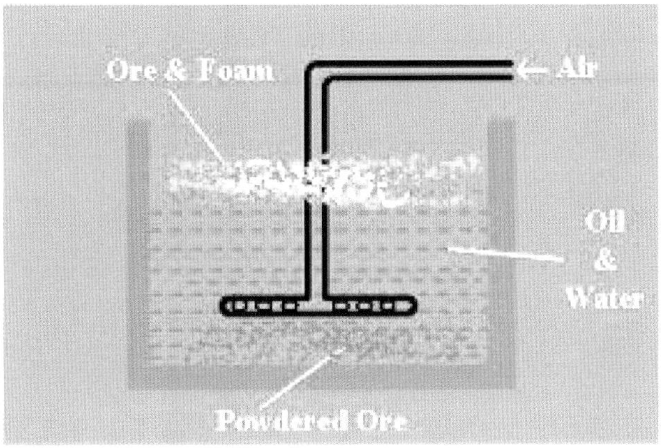

Roasting

The concentrated ore is then roasted in a furnace between $500^\circ C$ and $700^\circ C$ in the presence of a current of air. Sulphur is oxidized to SO_2 and impurities of As and Sb are removed as volatile oxides. The following reaction takes place.

$$2CuFeS_2 + O_2 \rightarrow Cu_2S + 2FeS + SO_2$$

$$S + O_2 \rightarrow SO_2$$

$$4As + 3O_2 \rightarrow As_2O_3$$

$$4Sb + 3O_2 \rightarrow 2Sb_2O_3$$

Cuprous sulphide and ferrous sulphide are further oxidized into their oxides.

$$2Cu_2S+3O_2 \rightarrow 2CU_2O+2SO_2$$

$$2FeS+3O_2 \rightarrow 2FeO+2SO_2$$

Smelting

The roasted ore is mixed with coke and silica (sand) SiO_2 and is introduced in to a blast furnace with $1200^{\circ}C$. The hot air is blasted and FeO is converted in to ferrous silicate $(FeSiO_3)$.

$$FeO + SiO_2 \rightarrow FeSiO_3$$

$$Cu_2O + FeS \rightarrow Cu_2S + FeO$$

$FeSiO_3$ (slag) floats over the molten matte of copper.

A matte of 60%Cu containing has been produced.

Bessemerization

Copper metal is extracted from molten matte through bessemerization. The matte is introduced in to Bessemer converter which uphold by tuyeres. The air is blown through the molten matte. Blast of air converts Cu_2S partly into Cu_2O which reacts with remaining Cu_2S to give molten copper.

$$2Cu_2S + 3O_2 \rightarrow 2Cu_2O + 2SO_2$$

$$2Cu_2O + Cu_2S \rightarrow 6Cu + SO_2$$

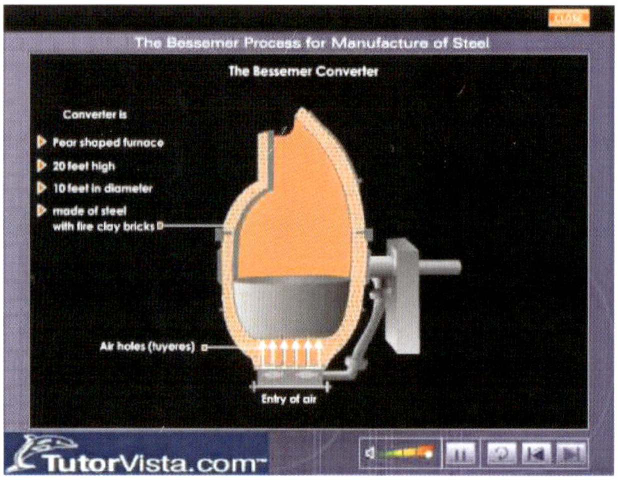

The copper so obtained is called "Blister copper" because, as it solidifies, SO_2 hidden in it escapes out producing blister on its surface.

Impurities in Blister

Copper and their Effects

Blister copper is 99% pure. It contains impurities mainly iron but little amount of As, Zn, Pb, Ag and Au may also be present. These impurities adversely affect the electrical as well as mechanical properties of copper. Therefore, they must be removed.

Refining of Copper

Blister copper is refined by electrolysis. Blocks of blister copper are casted to use as anodes and thin sheets of pure copper act as cathodes. The cathode plates are coated with graphite in order to remove depositing copper. The electrolyte is copper sulphate ($CuSO_4$) mixed with a little amount of H_2SO_4 to increase the electrical conductivity. Optimum potential difference is 1.3 volt for this electrolytic process. During electrolysis, pure copper (99.99%Cu) is deposited on the cathode plates and impurities which are soluble and fall to the bottom of the cell as anode mud or sludge.

Electrochemical Changes

During Electrolysis

$$Cu \rightarrow Cu^{+2} + 2e^{-} \text{(at the anode)}$$

$$Cu^{+2} + 2e^{-} \rightarrow Cu \text{ (at the cathode)}$$

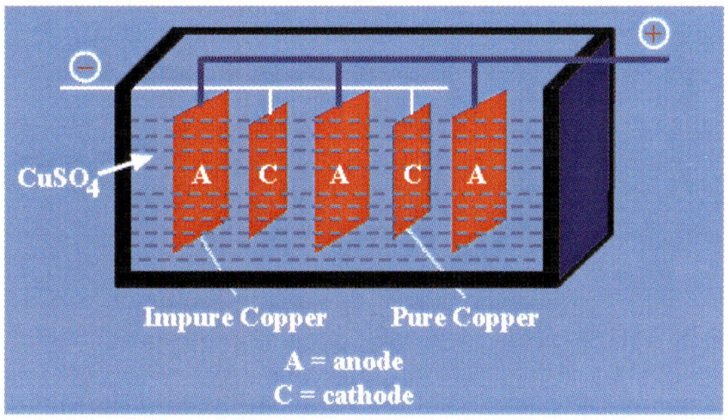

This electrically refined copper is 100% pure.

Gold Extraction

Even though the precious metal gold only occurs very rarely indeed, it can be found almost everywhere on the planet in the 16 kilometre thick Earth's crust. But the amount of gold in the earth is extremely small, so that mining is not really worthwhile. Generally there are two types of gold: primary gold ("mountain gold" from mines) and secondary gold ("stream gold" , "nuggets"). Above all the extraction of primary gold is associated with considerable effort. There are various methods to free the gold particles from their solid mantle in the ore extracted underground. The most important of these will be described here.

Extraction of Gold by Cyanide Leaching

The process of cyanide leaching is mainly used where the earth has a high gold concentration making an industrial development worthwhile. For this process the gold-bearing rock is crushed to sand and dust. Subsequently, the rock dust – piled in heaps, columns or stored in tanks – is combined with a sodium cyanide solution (sodium salt of hydrogen cyanide acid HCN). The acid detaches the gold from the rock dust and transports it in the acid seepage fluid in a chemically bound form. Subsequently, the gold is filtered out of the leach solution by adding zinc dust., then washed out of the muddy substrate and finally dried. The recovered raw gold is later refined into fine gold. Major reservations exist against this method of extraction due to the release of highly poisonous, easily flammable hydrogen cyanide acid, which presents significant risks for human beings and the environment.

Extraction of Gold by the Amalgam Process

The amalgam process is the oldest technical method of extracting gold and was used in antiquity. With the amalgam process the gold bearing rock is likewise crushed to fine sand. Afterwards mercury is added to the rock dust. Gold has the property that enables it to join to the surface of the mercury. The gold enriched mercury forms a silvery shining alloy solution, the so-called amalgam. The amalgam collects at the bottom of the mixing container and is easily separated from the other minerals. Subsequently, the amalgam is heated until the mercury evaporates leaving the pure raw gold. This method also involves dangers to health and the environment experienced when vaporizing highly poisonous mercury.

Extraction of Gold using the Electrowinning Process

Elecrowinning is in the meantime a frequently used process whereby gold is extracted without the use of toxic substances. This method uses an electrochemical approach. Here advantage is taken during the extraction of other precious metals such as f.e. copper or zinc were also small quantities of gold are found in the extracted rock. In the process of the electrolytic cleaning of the rock sludge the anode activates the precious metal bearing rock sludge. The cathode is made of a piece of pure metal. During electrolysis, the gold in the raw rock material is not dissolved and collects as a sludge under the anode. Together with gold other precious metals (f.e. Silver, copper, lead, tin) concentrate in this anode-sludge and can be easily skimmed off. Then using a suitable process the individual precious metals can be separated from each other.

Extraction of Gold using the Borax Process

The borax process is also an environmentally-friendly method of extracting gold. Here the gold-bearing rock material is placed in a melting crucible and then borax (sodium borate) is added. The borax sinks the melting point of the ore mixture and makes the molten mass more fluid. The low melting point thus enables gold to be extracted using a cost effective, low power heat sources to bring about melting. The gold sinks to the bottom of the crucible, whereas other rock material such as silicon, quartz, ore or generally all oxides rise to the top.

Silver Extraction

The extraction of silver by the solving processes simple. The ore is first roasted with salt in the usual way, whereby the formation of base metal chlorides cannot be avoided entirely. After roasting, the ore is first subjected to leaching with water, in order to extract the base metal chlorides, and then with hyposulphite of lime, to extract the silver.

Extraction of Silver

After a chloridizing roasting the ore should be examined to ascertain the amount of chloride of silver contained in it. In case the extraction should not be satisfactory, it is then easier to and what the cause is. The ore is then prepared for leaching.

First Leaching

The roasted ore contains chloride of silver, which does not dissolve in water, but generally there are also base chlorides in it, as the chlorides of copper, zinc, lead, iron, antimony, etc., which are soluble. It is the purpose of the first leaching to extract these base metals by means of hot water. For this purpose the ore is introduced into a tub or square box of pine wood, the planks being one and one-half to two inches thick.

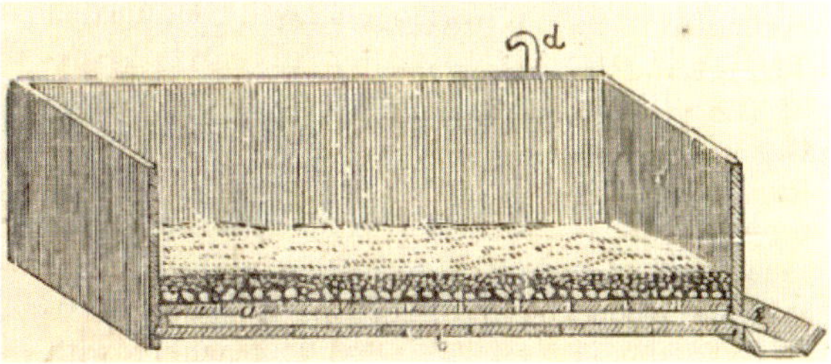

The boxes must be made as water-tight as possible and provided with a filter at the bottom. The filter is prepared in two ways, either as represented in figure a by fixing a false bottom, a, provided

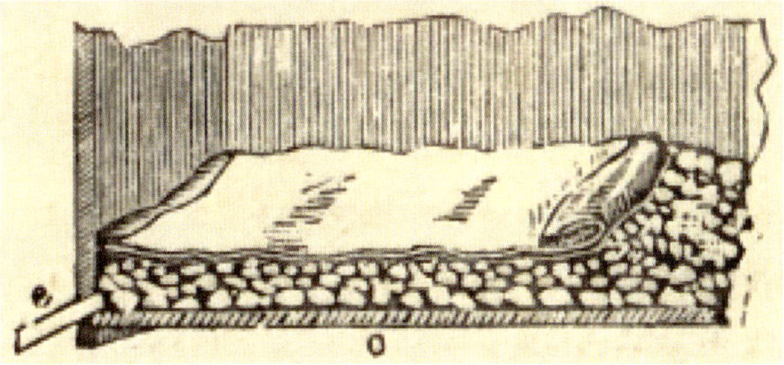

with numerous holes, one-half inch in diameter, about one inch above the bottom, b, or as figure b shows, without a false bottom. On the bottom, a, is thrown clean rock, quartz or poor ore of about the size of a hen's egg, three or four inches high; on this smaller stuff, and finally sand, free from mud. In figure b, rock of about the same size is thrown directly on the bottom, c, spread four inches high, then a few buckets full of rock, not smaller than hazel nuts, and only so much of it taken as to equalize the surface. This is then covered with a piece of canvas and is ready for use. The boxes, according to their size and the weight of the ore, may contain from one to five tons. Generally the ore must not be over fourteen or sixteen inches deep, but some ore allows a good leaching with twenty-five inches.

The roasted ore, generally without sifting, is charged into the boxes, and the surface spread evenly, leaving about six inches space from the top for the reception of the leaching water. The hotter the water is, the sooner it dissolves the soluble salts and the quicker the leaching progresses. It is conveyed through the pipe, d, and falls on a piece of canvas, whence it spreads equally and gently over the ore. The water soon reaches the bottom and begins to flow out through the pipe, e, into the trough, f.

In the beginning, the leaching water at e is highly charged with base metal salts, and shows a green color if there is much copper in the ore. The water is kept running in a continual stream till it reaches nearly to the rim of the box, when the influx and the efflux are equalized. After one or two hours a glass full of the liquid, at the pipe, e, is taken, and a few drops of sulphide of calcium (or of sodium) added. If a precipitate falls, of a dark or light color, the leaching must continue ; but it is not necessary to continue until no precipitate at all is perceived, as it requires some time— perhaps an hour—before all the water runs out after the pipe, d, is closed. The water which comes out last must be free from salts. This first leaching takes from two to four hours, sometimes longer.

Second Leaching

As soon as the ore is freed from the base chlorides soluble in water, a solution of hyposulphite of lime is led in from a tub or tank, on the ore, in order to dissolve the chloride of silver. This leaching is conducted like the former. It depends on the amount of silver how long this work continues—from eight to twenty hours. The clear cold solution, containing the chloride of silver in the form of a double salt, has a very sweet taste, and is conveyed through a trough or Indian rubber hose into a precipitating tub. Very rich ore, containing 12 to 15 per cent, of silver, would require forty-eight hours leaching, and even then it would be necessary to subject the ore to a second leaching with the hyposulphite, with an intermediate roasting with green vitriol and salt; for, with the best work, if 95 per cent, are extracted, the tailings would still appear sufficiently rich for this, containing about 200 ounces of silver per ton. Ores containing $350 per ton are often leached out perfectly in twelve hours. The end of the lixiviation is ascertained in the same way as in leaching with water, using the sulphide of calcium. If no precipitate is obtained the extraction is finished.

The color of the precipitate is a black-brown. The presence of base metals changes the color somewhat. Iron makes it black; copper, red- brown; lead and antimony, light red-brown, etc. The silver is first dissolved, especially if a diluted solution of hyposulphite of lime is used; and for this reason the first precipitate is the richest in silver. Ore containing a great deal of lead— especially if the roasting was so conducted that a large part of it remained as sulphate of lead, which is not soluble in the leaching water—will give in the beginning of the leaching with the solvent a precipitate of silver with some lead; afterwards, however, the silver diminishes, so that the precipitate of lead finally appears free of silver. Besides the sulphate of lead, sub-chlorides and oxy-chlorides are formed during the roasting which are not soluble in water, but are dissolved by the hyposulphite of lime; for this reason always some base metals will be found in the precipitate.

In case rebellious ores are treated, and hot water is used for the extraction of base chlorides, a better silver is obtained if the ore is cooled down by cold water before the cold and diluted solvent is applied. Purer ores may be treated with a warm solution of the solvent.

When examining the tailings as to the amount of silver left therein, it must be remembered that; after leaching out a quantity of metals by water and the solvent, the ore lost a considerable part of its original weight, and that consequently one-half ounce of such tailings taken into assay will always give a larger silver button than there ought to be. A true assay of leached tailings is made if half an ounce of the same ore is leached on a filter with hot water and hyposulphite of lime, in the same way as the ore on a large scale, washed with water, dried and weighed. The weight found after leaching must be taken for half an ounce in assaying the tailings.

The residue, or tailings in the leaching box, must be removed now as valueless. The sides of the leaching boxes are from eighteen to twenty-four inches above the bottom, and being from six to eight feet square in the clear, the removing of the tailings by means of chloride is easily effected. The men must be careful not to dig too deep otherwise the filter will be injured. It is quite proper to x wooden staves, as long as the box requires, on top of the filter. These staves are one inch wide and one-half of an inch thick, and are placed from four to five inches apart, so as to protect the canvas or filter against the shovel. In figure the staves are laid upon the canvas.

The leaching boxes or tubs may be arranged so that, being tipped over, the whole charge falls out at once. In this case the filter must be made in a different way from that described above. Rocks are not serviceable here. On the false bottom, a, of figure a, a layer of thin, leafless switches is placed, and on this another one crosswise, then covered with a piece of canvas, and secured with some staves to prevent the falling out of the whole filter when turned over.

Precipitation of the Silver

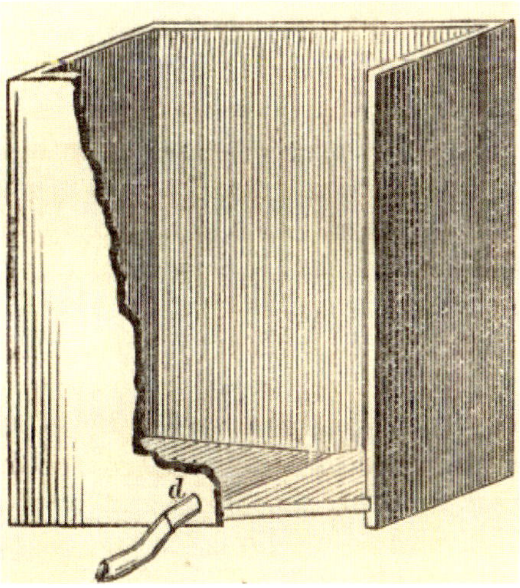

The liquid of the second leaching is conveyed through a trough or India rubber hose to the precipitating tanks, of which three or four are employed. If tubs are used, which for this purpose are the best, they are from three to four feet in diameter, and four feet high. The tanks or boxes have a rectangular shape of about the same capacity, the bottom being inclined toward the middle, as shown by figure. The hyposulphite of lime, as it comes from the leaching tanks, is conducted into these until they are more than two-thirds full. The trough or hose is then changed to discharge the liquid into the next precipitating tub, while the precipitation of the first commences.

The liquid used for precipitating the silver is sulphide of calcium. It is poured in until all the silver is supposed to be precipitated, and at the same time the solution is stirred vigorously. Treating always the same kind of ore, the required quantity of the precipitating agent is soon learned. The black precipitate sinks to the bottom, and the workman now dips a little of the clear liquid out in a glass tube or tumbler, and adds a few drops of the sulphide of lime. If a dark precipitate or a dark color is produced, it shows that there is still silver in the liquid, and more of the agent must be added; but if on the contrary no precipitate is observed, there is either enough or too much of the sulphide. To prove this, some of the silver-holding liquid is added to a test, taken from the tank under treatment. If in this case a precipitate is formed, silver-holding liquid must be carefully added to the tank until no reaction is produced. This work, delicate as it seems, is easily learned by the workmen. If a little silver should be left in the liquid, it is not injurious, neither is the silver to be considered as lost, because the same liquid is used over again; but a small excess of the sulphide of calcium would cause a loss in silver, as it precipitates sulphide of silver in the leaching tank in the mass of ore, which is not dissolved again. The precipitation is performed in a short time, requiring about fifteen minutes for each tank. The stirring must be executed with vigor. Wooden grates fixed to a vertical stem will answer the purpose.

The clear solution above the settled precipitate is pumped or elevated to the reservoir, whence it was conveyed on the ore. It is now ready to be used again. The sulphide of calcium having performed its duty in precipitating the silver, is turned into hyposulphite of lime, thus replacing all of the solvent.

To prevent small floating particles of silver from being elevated with the liquid, it is well to allow sufficient time for the precipitated silver to settle. For this reason it is better to have more precipitating tanks or tubs. It is not necessary to remove the silver after each precipitation. The clear liquid can be drawn off, by means of a syphon, from all the precipitating tubs into a general receiver, whence it may be pumped up. After the solvent has been removed, the precipitated silver can be drawn off through the pipe, d, figure, directly into canvas bags.

Treatment of the Precipitated Silver

The black precipitate of sulphide of silver is conveyed directly into filters made of canvas, either in the shape of pointed bags, like those used for amalgam, or in the shape of common bags. As soon as all the liquid runs out, pure water (if possible, warm) is poured on the silver, and this repeated several times till no taste is observed in the filtering water. The precipitate, while still in the bags, is placed beneath a screw press and the fluid pressed out as completely as possible. The black silver cakes are then taken out and dried in a warm room or in a drying oven. For the purpose of burning off the sulphur, the dried sulphide is introduced into a muffle or other calcining furnace, and heated till the sulphur commences to burn with its known blue flame. When this disappears the heating must continue at a dark red heat for one or two hours. By this operation the cakes are reduced almost entirely to metallic silver, generally covered with threads of silver; sometimes an intense green color is assumed by pieces remaining in the furnaces overnight.

The burned cakes are now prepared for smelting in crucibles. They are placed in black lead crucibles, according to the size, up to three hundred pounds, and fused. All the sulphur was not driven out by the preceding operation. The remaining part must be removed by placing metallic iron (5, d) in the fused metal; thereby iron matt is formed, which rises to the surface and is skimmed off. The surface of the silver is then cleaned by adding some bone ash and borax, or borax alone, which is also skimmed off and the silver dipped out or poured out into moulds. According to the careful treatment in the roasting process, and the nature of the ore, the silver will be from 800 to 950 ne.

Mr. O. Hofmann, in need of sulphur for the production of sulphide of calcium, used to calcine the dried sulphide of silver in iron retorts. In this way he obtained a large proportion of sulphur as a fine sublimate. This could be done also in a proper muffle furnace, so arranged that after all obtainable sulphur had sublimated in a receiver this could be removed and the calcination continued under access of air.

Quality of Ores fit for the Solving Process

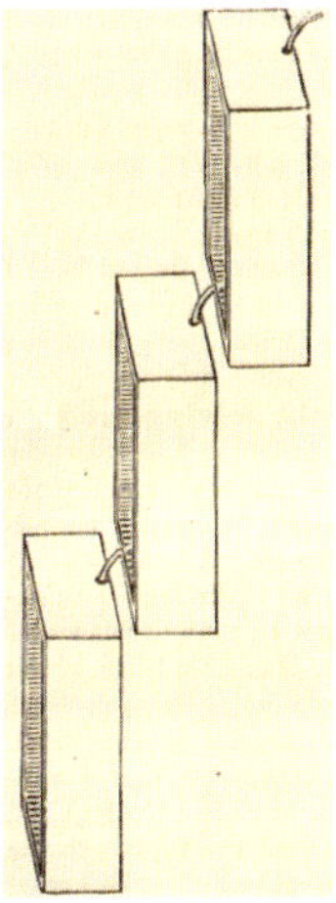

There is no process so suitable for all kinds of ores as the solving process. Generally considered, all silver ores can be treated by the solving process which are subjected to the pan amalgamation after roasting ; but in many instances— especially with the rebellious ores— a better result is obtained by this than by working in pans. The great advantage of this process is cheapness. Roasting of course is indispensable except with chloride ores; but neither pans and the required power, nor quicksilver, are used, and for this reason less capital is necessary to put up reduction works. All the cupreous silver ores of Cerro Gordo, Yellow Pine, Montgomery, and of the other now silver districts, can be treated to great advantage by the solving process.

Two objections have to be considered. First, there is more water required than for pan amalgamation—at least this is the case with rebellious ores; but the quantity of water depends on the quantity of base metals in the ore, and also on the arrangement of the leaching boxes. One box, containing one ton of ore, requiring three hours leaching, may consume 250 gallons of water; three boxes of the same size would take three times as much water if placed on the same level; but by arranging the boxes in a less favorable position, one above the other, as shown in figure, only one-half of the quantity of water is needed. It takes more time to leach three boxes together than a single one. Leaching the ore with hyposulphite of lime, the supply of this in the first box must be stopped if no more silver comes out, and the solution carried, by means of hose, to the second, and then to the third box, if it should be necessary; so that while the second box is yet under leaching, the first can be discharged and a new charge introduced. All three boxes should receive clean water at the start. This arrangement should be adopted only when rendered necessary by the scarcity of water. To have all leaching boxes on a level is preferable.

The other objection is conned to a certain class of ores containing clay and lime. If pulverized, so much ne pulp will be produced that the leaching is impossible. It is not advisable to crush the ore coarser than will allow of its passing through a sieve of forty holes to the inch, in some cases, perhaps, through thirty-five holes; and if with such crushing a fine clay pulp is produced, the ore is unfit for all leaching processes, unless wet crushing is adopted, in order to separate the slime from the sand, as Mr. O. Hofmann was compelled to arrange in Trinidad, Sonora. In this case, a separate drying for the purpose of roasting is not necessary if long furnaces are in use. It is not unlikely that for similar ore and pan tailings an agitating filtering box could be constructed which would render the leaching possible.

Sulphide of Calcium

Sulphide of calcium for the precipitation of silver is preferable to the sulphide of sodium, principally for the reason that its manufacture is cheaper and more easy, but also on account of the quality of the precipitated silver, which is easier to wash, to press and to desulphurize. The sulphide of calcium is easily obtained and manufactured on the ground where the mill is situated. The articles required for this purpose are brimstone (worth about four cents per pound) and burned lime. The sulphide is formed only from caustic lime, consequently more is obtained from fresh burned lime. Of this a certain quantity is charged into an iron kettle, water added, and then the pulverized sulphur. The proportion of sulphur and lime depends on the quality of the latter. The purest quality of lime from Santa Cruz, Cal., for instance, takes one pound of sulphur to 1.33 of lime. Of poorer qualities of lime it is better to take three pounds to one of sulphur and about ten parts of water. It is kept boiling for two or three hours, stirred with wooden shovels, and then discharged into a filtering box, prepared like figure. The clear, dark, yellow-red sulphide of calcium comes out from below the filter, and can be kept in iron vessels. The liquid ought to be between 5° and 6° Beaume. The residue is washed with water, whereby a diluted uid is obtained, which is used with the lime of the next charge. Mr. E. Smyth, in La Dura, Mexico, treats the lime and sulphur with steam. This has the advantage of dispensing with the stirring, and may be performed also in wooden vessels. The steam replaces the fire and has no chemical influence on the quality of the sulphide.

Zinc Extraction

Zinc is found in the earth's crust primarily as zinc sulfide (ZnS). Rarely is the ore, as mined, rich enough to be used directly by smelters; it needs to be concentrated. Zinc ores contain 5-15% zinc. To concentrate the ore, it is first crushed and then ground to enable optimal separation from the other minerals. Typically, a zinc concentrate contains about 55% of zinc with some copper, lead and iron. Zinc concentration is usually done at the mine site to keep transport costs to smelters as low as possible.

Zinc Smelting

Zinc smelting is the process of converting zinc concentrates (ores that contain zinc) into pure zinc. The most common zinc concentrate processed is zinc sulfide, which is obtained by concentrating sphalerite using the froth flotation method. Secondary (recycled) zinc material, such as zinc oxide, is also processed with the zinc sulfide. Approximately 30% of all zinc produced is from recycled sources.

There are two methods of smelting zinc: the pyrometallurgical process and the hydrometallurgical process (over 90% of hydrometallurgical process is in electrolytic plants using electrolysis process). Both methods are still used. Both of these processes have the same first step: roasting.

Roasting is a process of oxidizing zinc sulfide concentrates at high temperatures into an impure zinc oxide, called "Zinc Calcine". The chemical reactions taking place during the process are:

$$2ZnS + 3O_2 \rightarrow 2ZnO + 2SO_2$$

$$2SO_2 + O_2 \rightarrow 2SO_3$$

The process of roasting varies based on the type of roaster used. There are three types of roasters: multiple-hearth, suspension, and fluidized-bed.

Hydrometallurgical Process

Today over 90% zinc is produced hydrometallurgically in electrolytic plants.

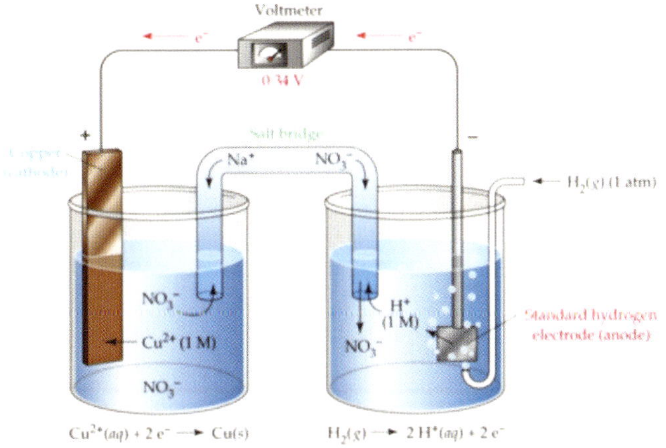

Electrolysis Process

The electrolysis process, also known as the hydrometallurgical process, Roast-Leach-Electrowin (RLE) process, or electrolytic process, is more widely used than the pyrometallurgical processes.

The electrolysis process consists of 4 steps: leaching, purification, electrolysis, and melting and casting.

Leaching

The basic leaching chemical formula that drives this process is:

$$ZnO + SO_3 \rightarrow ZnSO_4$$

Purification

The purification process utilizes the cementation process to further purify the zinc. It uses zinc dust and steam to remove copper, cadmium, cobalt, and nickel, which would interfere with the electrolysis process.

Electrolysis

Zinc is extracted from the purified zinc sulfate solution by electrowinning, which is a specialized form of electrolysis. The process works by passing an electric current through the solution in a series of cells. This causes the zinc to deposit on the cathodes (aluminium sheets) and oxygen to form at the anodes. Sulfuric acid is also formed in the process and reused in the leaching process. Every 24 to 48 hours, each cell is shut down, the zinc-coated cathodes are removed and rinsed, and the zinc is mechanically stripped from the aluminium plates.

Electrolytic zinc smelters contain as many as several hundred cells. A portion of the electrical energy is converted into heat, which increases the temperature of the electrolyte. Electrolytic cells operate at temperature ranges from 30 to 35 °C (86 to 95 °F) and at atmospheric pressure. A portion of the electrolyte is continuously circulated through the cooling towers both to cool and concentrate the electrolyte through evaporation of water. The cooled and concentrated electrolyte is then recycled to the cells. This process accounts for approximately one-third of all the energy usage when smelting zinc.

There are two common processes for electrowinning the metal: the low current density process, and the Tainton high current density process. The former uses a 10% sulfuric acid solution as the electrolyte, with current density of 270–325 amperes per square meter. The latter uses 22–28% sulfuric acid solution as the electrolyte with a current density of about 1,000 amperes per square metre. The latter gives better purity and has higher production capacity per volume of electrolyte, but has the disadvantage of running hotter and being more corrosive to the vessel in which it is done. In either of the electrolytic processes, each metric ton of zinc production expends about 3,900 kW•h (14 GJ) of electric power.

Melting and Casting

The final step is to melt the cathodes in an induction furnace. It is then either cast into pure zinc (99.995% pure) ingots or alloyed and cast into ingots.

Pyrometallurgical Process

There are also several pyrometallurgical processes that reduce zinc oxide using carbon, then distil the metallic zinc from the resulting mix in an atmosphere of carbon monoxide. The major downfall of any of the pyrometallurgical process is that it is only 98% pure; a standard composition is 1.3% lead, 0.2% cadmium, 0.03% iron, and 98.5% zinc. This may be pure enough for galvanization, but not enough for die casting alloys, which requires special high-grade zinc (99.995% pure). In order to reach this purity the zinc must be refined. The four types of commercial pyrometallurgical processes are the St. Joseph Minerals Corporation's (electrothermic) process, the blast furnace-process, the New Jersey Zinc continuous vertical-retort process, and the Belgian-type horizontal retort process.

St. Joseph Mineral Company (Electrothermic) Process

This process was developed by the St. Joseph Mineral Company in 1930, and is the only pyrometallurgical process still used in the US to smelt zinc. The advantage of this system is that it is able to smelt a wide variety of zinc-bearing materials, including electric arc furnace dust. The disadvantage of this process is that it is less efficient than the electrolysis process.

The process begins with a downdraft sintering operation. The sinter, which is a mixture of roaster calcine and EAF calcine, is loaded onto a gate type conveyor and then combustions gases are pumped through the sinter. The carbon in the combustion gases react with some the impurities, such as lead, cadmium, and halides. These impurities are driven off into filtration bags. The sinter after this process, called product sinter, usually has a composition of 48% zinc, 8% iron, 5% aluminium, 4% silicon, 2.5% calcium, and smaller quantities of magnesium, lead, and other metals.

The sinter product is then charged with coke into an electric retort furnace. A pair of graphite electrodes from the top and bottom furnace produce current flow through the mixture. The coke provides electrical resistance to the mixture in order to heat the mixture to 1,400 °C (2,550 °F) and produce carbon monoxide. These conditions allow for the following chemical reaction to occur:

$$ZnO + CO \rightarrow Zn(vapour) + CO_2$$

The zinc vapor and carbon dioxide pass to a vacuum condenser, where zinc is recovered by bubbling through a molten zinc bath. Over 95% of the zinc vapor leaving the retort is condensed to liquid zinc. The carbon dioxide is regenerated with carbon, and the carbon monoxide is recycled back to the retort furnace.

Blast Furnace Process

This process was developed by the National Smelting Company at Avonmouth Docks, England, in order to increase production, increase efficiency, and decrease labor and maintenance costs. L. J. Derham proposed using a spray of molten lead droplets to rapidly chill and absorb the zinc vapor, despite the high concentration of carbon dioxide. The mixture is then cooled, where the zinc separates from the lead. The first plant using this design opened up in 1950. One of the advantages of this process is that it can co-produce lead bullion and copper dross. In 1990, it accounted for 12% of the world's zinc production.

The process starts by charging solid sinter and heated coke into the top of the blast furnace. Preheated air at 190 to 1,050 °C (370 to 1,920 °F) is blown into the bottom of the furnace. Zinc vapor and sulfides leave through the top and enter the condenser. Slag and lead collect at the bottom of the furnace and are tapped off regularly. The zinc is scrubbed from the vapor in the condenser via liquid lead. The liquid zinc is separated from the lead in the cooling circuit. Approximately 5,000 metric tons (5,500 short tons) of lead are required each year for this process, however this process recovers 25% more lead from the starting ores than other processes.

The IS process is an energy-intensive process and thus became very expensive following the rise of energy prices. Today, Imperial Smelting furnaces are only in operation in China, India, Japan and Poland.

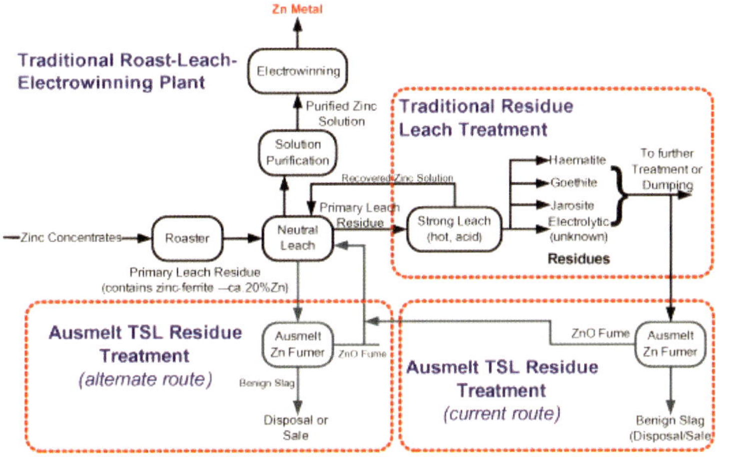

New Jersey Zinc continuous vertical retort

New Jersey Zinc Continuous Vertical Retort

The New Jersey Zinc process is no longer used to produce primary zinc in the U.S., Europe and Japan, however it still is used to treat secondary operations. This process peaked in 1960, when it accounted for 5% of world zinc production. A modified version of this process is still used at Huludao, China, which produced 65,000 metric tons per year.

This process begins by roasting concentrates that are mixed with coal and briquetted in two stages. The briquettes are then heated in anautogenous coker at 700 °C (1,292 °F) and then charged into the retort. There are three reasons to briquette the calcine: to ensure free downward movement of the charge; to permit heat transfer across a practical size cross-section; to allow adequate porosity for the passage of reduced zinc vapor to the top of the retort. The reduced zinc vapor that is collected at the top of the retort is then condensed to a liquid.

Overpelt improved upon this design by using only one large condensation chamber, instead of many small ones, as it was originally designed. This allowed for the carbon monoxide to be recirculated into the furnaces for heating the retorts.

This process was licensed to the Imperial Smelting Company (ISC), in Avonmouth, England, which had a large vertical retort (VR) plant in production for many years. It was used until the mid-1970s when it was superseded by the company's Imperial Smelting Furnace (ISF) plant. The VR plant was demolished in 1975.

Belgian-type Horizontal Retort Process

This process was the main process used in Britain from the mid-19th century until 1951. The process was very inefficient as it was designed as a small scale batch operation. Each retort only produced 40 kilograms (88 lb), so companies would put them together in banks and used one large gas burner to heat all of them. The Belgian process requires redistillation to remove impurities of lead, cadmium, iron, copper, and arsenic.

Sodium Extraction

Sodium is the seventh most abundant element by weight found in the earth's crust. The largest source of sodium is sodium chloride (NaCl) which occurs as rock salt and in sea water.

Castner's Process

- Albite (soda feldspar) $NaAlSi_3O_8$
- Borax (as tincal) $Na_2B_2O_7.10H_2O$
- Glauber's salt $Na_2SO_4.10H_2O$
- Sodium chloride, (common salt) found as rock salt, in sea water and in lakes.
- Sodium nitrate ($NaNO_3$) as chile saltpetre.
- Extraction of Sodium

  The metal sodium, was prepared in 1807 by an English chemist,

Sir Humphery Davy by electrolysis of fused caustic soda. The two methods used for extracting sodium metals are,

- Castner's process: By the electrolysis of fused caustic soda.
- Down's process: By the electrolysis of fused sodium chloride.

- Castner's process

  Molten sodium hydroxide (caustic soda, NaOH) is electrolyzed using an iron cathode and nickel anode. A nickel wire gauze cylinder prevents the electrodes from touching each other. On passing electric current through the melt, sodium is liberated at the cathode and oxygen at the anode. The liberated sodium metal floats over the surface (lighter than fused caustic soda), and collects inside the receiving vessel where hydrogen prevents sodium from oxidation. Excess of the gas escapes from the outlet. Sodium is removed from time to time with the help of perforated spoons and kept under kerosene oil. The various reactions taking place during electrolysis are:

  $NaOH \underline{\text{melting}}\, Na^{+} + OH^{-} \quad (\text{dissoiation})$

  At cathode:

  $Na^{+} + e^{-} \rightarrow Na$

  At anode:

  $OH^{-} \rightarrow \frac{1}{2}H_2O + \frac{1}{4}O_2 + 2e^{-}$

Most of the water formed in the reaction gets evaporated, and the rest is electrolyzed into hydrogen and oxygen. Therefore, metallic sodium and hydrogen are liberated at the cathode and oxygen at the anode.

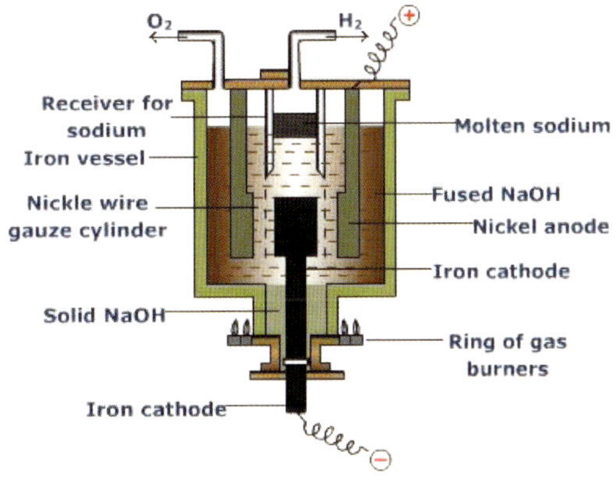

Figure: Castner's process

Downs Process for Extraction of Sodium

In this method sodium is obtained by the electrolysis of a mixture of sodium chloride (40%) and calcium chloride (60%).The $CaCl_2$ is added to lower the melting point of sodium chloride from 1085 to 850K.

The reasons for lowering the temperature are:

i) The melting point of NaCl is very high and it is very difficult to maintain this temperature.

ii) Sodium is volatile at high temperature and therefore, a part of the metal formed may vaporize forming a fog.

iii) Chlorine produced as by product may corrode the material of the vessel at this temperature.

iv) The metal obtained at this temperature is in colloidal state and its separation will be difficult.

The cell consists of steel tank with heat resistant bricks. A circular graphite anode is placed in the center of the cell. It is surrounded by cylindrical iron cathode. A dome shaped steel hood covers the anode which provides outlet for the escape of Cl_2 gas.

A fine wire gauze diaphragm separates the anode and cathode through which molten NaCl may pass but molten sodium cannot pass. The wire gauze separates anode compartments so that the sodium and chlorine formed are not able to intermix and react with each other.

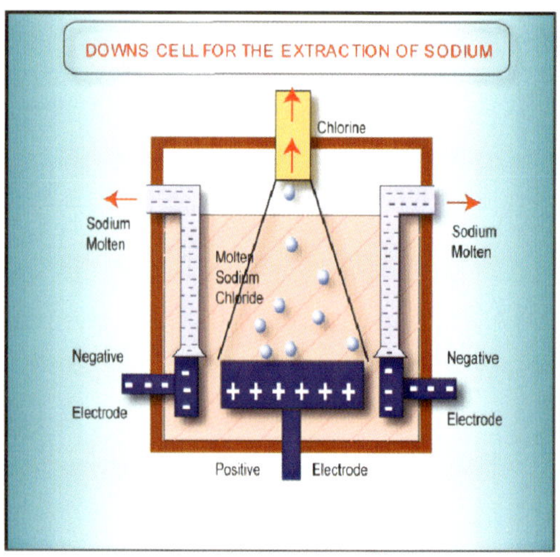

DOWNS CELL FOR THE EXTRACTION OF SODIUM

The sodium produced at cathode is in the molten state. Since it is lighter than the molten electrolyte, it rises to the surface and taken out from the outlet. The following reactions occur during electrolysis.

$$NaCl \longrightarrow \underset{Cathode}{Na^+} + \underset{Anode}{Cl^-}$$

At cathode $Na^+ + e^- \longrightarrow Na$

At anode $Cl^- - e^- \longrightarrow Cl$

$Cl + Cl \longrightarrow Cl_2$

Sodium obtained by this method is 99.5% pure.

Advantage of Down's Process

1) Sodium metal obtained has high degree of purity (99.5%).
2) Starting material i.e. sodium chloride is very cheap.
3) Chlorine obtained is useful by product.

Magnesium Extraction

Magnesium is found in solution in sea-water (about 1.3 kg m-3 magnesium) and in natural brines. It is also found extensively in the ores magnesite ($MgCO_3$) and dolomite ($MgCO_3.CaCO_3$). Both of the two types of resources can be used as raw materials to manufacture magnesium metal.

Magnesium is principally produced by two methods:

a) Thermal reduction of magnesium oxide.
b) Electrolysis of magnesium chloride.

Thermal Reduction Process

Dolomite is normally treated by the thermic method: this consists of reducing magnesium oxide, produced by calcining the dolomite raw material, with ferrosilicon to produce metallic magnesium and a calcium iron silicate slag.

Step 1: Dolomite ore is crushed and heated in a kiln to produce a mixture of magnesium and calcium oxides, a process known as calcining:

$$MgCO_3.CaCO_3(s) \rightarrow MgO.CaO(s) + 2CO_2(g)$$

Step 2: The next step is reduction of the magnesium oxide. The reducing agent is ferrosilicon (an alloy of iron and silicon) which is made by heating sand with coke and scraps iron, and typically contains about 80% silicon.

Step 3: The oxides are mixed with crushed ferrosilicon, and made into briquettes for loading into the reactor. Alumina may also be added to reduce the melting point of the slag. The reaction is carried out at 1500 - 1800 K under very low pressure, close to vacuum. Under these conditions the magnesium is produced as a vapor which is condensed by cooling to about 1100 K in steel-lined condensers, and then removed and cast into ingots:

$$2MgO(s) + Si(s) \rightleftharpoons SiO_2(s) + 2Mg(g)$$

Step 4: The forward reaction is endothermic and the position of equilibrium is in favor of magnesium oxide. However, by removing the magnesium vapor as it is produced, the reaction goes to completion. The silica combines with calcium oxide to form the molten slag, calcium silicate:

$$2MgO(s) + Si(s) \rightleftharpoons SiO_2(s) + 2Mg(g)$$

The process gives magnesium with up to 99.99% purity, slightly higher than from the electrolytic processes.

Electrolytic Process

Used to treat magnesite, serpentine, seawater magnesia and brines from carnallite and lake waters, the electrolytic process is usually the preferred choice outside China.

The process involves two stages:

i) Production of pure magnesium chloride from sea water or brine.

ii) Electrolysis of fused magnesium chloride.

i) Production of Pure Magnesium Chloride from Sea Water or Brine

Where sea-water is the raw material, it is treated with dolomite which has been converted to mixed oxides by heating to a high temperature. Magnesium hydroxide precipitates, while calcium hydroxide remains in solution. Magnesium hydroxide is filtered off and on heating readily forms the pure the oxide.

Conversion to magnesium chloride is achieved by heating the oxide, mixed with carbon, in a stream of chlorine at a high temperature in an electric furnace.

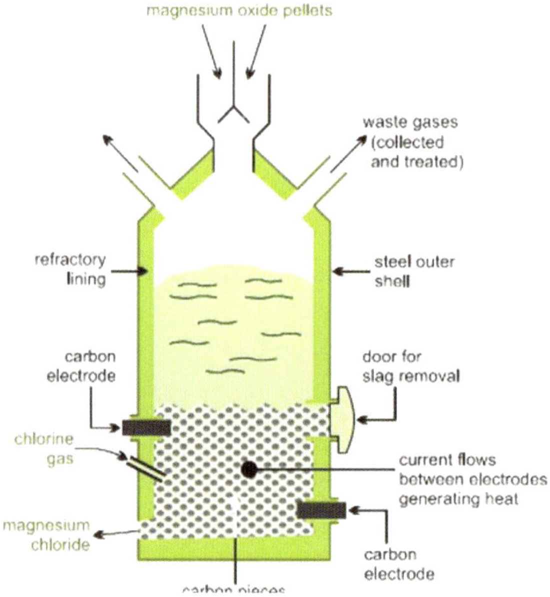

Figure illustrating the production of magnesium chloride from magnesium oxide

Several reactions occur:

$$2MgO(s) + C(s) + 2Cl_2(g) \rightarrow 2MgCl_2(s) + CO_2(g)$$
$$Cl_2(g) + C(s)\ H_2O(g) \rightarrow 2HCl(g) + CO(g)$$
$$MgO(g) + 2HCl(g) \rightarrow MgCl_2(s) + H_2O(g)$$

Where magnesium chloride-rich brines are the source of magnesium, the solution is treated for removal of various impurities and the remaining magnesium chloride solution concentrated by evaporation in several stages.

The last stage of dehydration has to be carried out in the presence of hydrogen chloride gas to avoid hydrolysis of the magnesium chloride:

$$2Mg(OH)Cl(s) + HCl(g) \rightarrow MgCl_2(s) + H_2O(g)$$

A new process is under development using magnesite. Small pieces of the ore are converted directly to molten magnesium chloride by heating with chlorine in an electric furnace in the presence of carbon monoxide.

ii) The Electrolysis of Fused Magnesium Chloride

The resulting anhydrous magnesium chloride is fed continuously into electrolytic cells which are hot enough to melt it.

On electrolysis, magnesium and chlorine are produced:

$$\oplus \text{ Anode}: \quad 2Cl^- \rightarrow Cl_2 + 2e^-$$

$$\ominus \text{Cathode} \quad Mg^{2+} + 2e^- \rightarrow Mg$$

The molten metal is removed and cast into ingots. The chlorine gas is recycled to the chlorination furnace.

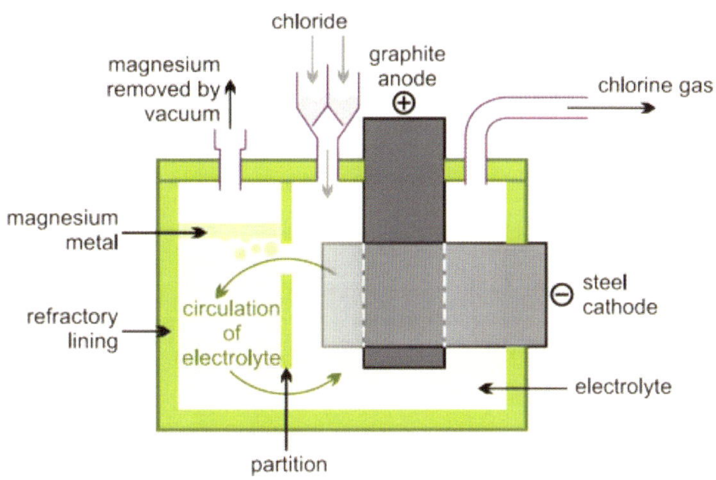

Figure illustrating the electrolysis of magnesium chloride

References

- Georg Müller (1988) Crystal growth from the melt Springer-Verlag, Science 138 pages ISBN 3-540-18603-4, ISBN 978-3-540-18603-8
- Concentration-of-ore, general-principles-and-processes-of-isolation-of-elements: toppr.com, Retrieved 16 April 2018
- Elizabeth R. Burkhardt "Potassium and Potassium Alloys" in Ullmann's Encyclopedia of Industrial Chemistry, Wiley-VCH, 2006. doi:10.1002/14356007.a22_031.pub2
- Magnetic-separation-method: 911metallurgist.com, Retrieved 25 June 2018

- Osborne, Graeme (1981). "Guillaume Daniel Delprat". Australian Dictionary of Biography. Canberra: Australian National University. Retrieved 7 June 2012.
- General-principles-and-processes-of-isolation-of-elements/extraction-of-crude-metal-from-concentrated-ore, chemistry: toppr.com, Retrieved 26 March 2018
- Malozemoff, Plato (March 1941). ""Operating Characteristics of Mechanical Flotation Machines,"". Engineering & Mining Journal: 45-49.
- What-is-liquation: wisegeek.com, Retrieved 27 April 2018
- Lynch, Martin (2002). Mining in World History. London: Reaktion Press. p. 208. ISBN 978-1-86189-173-0.
- Refining, general-principles-and-processes-of-isolation-of-elements, chemistry: toppr.com, Retrieved 09 April 2018
- Sebenik, Roger F. et al. (2005) "Molybdenum and Molybdenum Compounds" in Ullmann's Encyclopedia of Chemical Technology. Wiley-VCH, Weinheim. doi:10.1002/14356007.a16_655
- Extraction-Properties-and-Uses-of-Sodium-1157977: scribd.com, Retrieved 20 July 2018
- Voith EcoCell flotation plant "Archived copy" (PDF). Archived from the original (PDF)on 2009-08-24. Retrieved 2009-01-08.

4

Alloys

A combination of metals or metal and an element is called an alloy. The metals such as aluminium, iron, copper, zinc, chromium, etc. are often used for making alloys. The diverse topics elaborated in this chapter such as substitutional alloys, interstitial alloys, amalgam, amorphous metal, oxide dispersion strengthened (ODS) alloy, shape-memory alloy, etc. will provide a comprehensive understanding of the subject matter.

Alloys are metallic compounds composed of one metal and one or more metal or non-metal element.

Examples of common alloys:

- Steel, a combination of iron (metal) and carbon (non-metal)
- Bronze, a combination of copper (metal) and tin (metal) and
- Brass, a mixture of copper (metal) and zinc (metal)

Structure of Alloys

If we look at a metal through a powerful electron microscope, we can see the atoms inside arranged in a regular structure called a crystalline lattice. Imagine a small cardboard box full of marbles and that's pretty much what we'd see. In an alloy, apart from the atoms of the main metal, there are also atoms of the alloying agents dotted throughout the structure. (Imagine dropping a few plastic balls into the cardboard box so they arrange themselves randomly among the marbles).

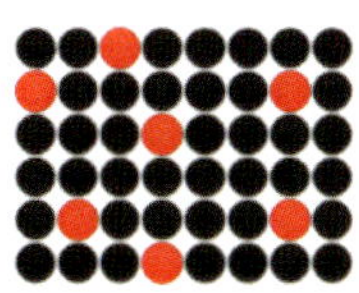

Substitution Alloy

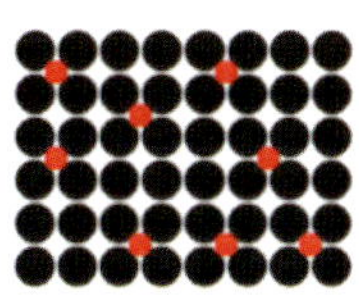

Interstitial Alloy

Substitution Alloys

If the atoms of the alloying agent replace atoms of the main metal, we get what's called a substitution alloy. An alloy like this will form only if the atoms of the base metal and those of the alloying agent are of roughly similar size. In most substitution alloys, the constituent elements are quite near one another in the periodic table. Brass, for example, is a substitution alloy based on copper in which atoms of zinc replace 10–35 percent of the atoms that would normally be in copper. Brass works as an alloy because copper and zinc are close to one another in the periodic table and have atoms of roughly similar size.

Interstitial Alloys

Alloys can also form if the alloying agent or agents have atoms that are very much smaller than those of the main metal. In that case, the agent atoms slip in between the main metal atoms (in the gaps or "interstices"), giving what's called an interstitial alloy. Steel is an example of an interstitial alloy in which a relatively small number of carbon atoms slip in the gaps between the huge atoms in a crystalline lattice of iron.

Properties

Individual pure metals may possess useful properties such as good electrical conductivity, high strength, and hardness, or heat and corrosion resistance. Commercial metal alloys attempt to combine these beneficial properties in order to create metals more useful for particular applications than any of their component elements.

Steel, for example, requires the right combination of carbon and iron (about 99% iron and 1% carbon, as it turns out) in order to produce a metal stronger, lighter and more workable metal than pure iron.

Precise properties of new alloys are difficult to calculate because elements do not just combine to become a sum of parts; rather, they form through chemical interactions which depend upon component parts and specific production methods. As a result, much testing is required in the development of new metal alloys.

Melting temperature is a key factor in alloying metals. Galinstan, a low-melt alloy containing gallium, tin, and indium, is liquid at temperatures above 2.2°F (-19°C), meaning that its melting point is 122°F (50°C) lower than pure gallium and more than 212°F (100°C) below indium and tin.

Galinstan and Wood's Metal are examples of eutectic alloys - alloys having the lowest melting point of any alloy combination containing the same elements.

Composition

Thousands of alloy compositions are in regular production with new compositions being developed every year.

Accepted standard compositions include the purity levels of constituent elements (based on weight content). The makeup, as well as mechanical and physical properties of common alloys, are standardized by international organizations such ISO, SAE International, and ASTM International.

Production

Some metal alloys are naturally occurring and require little processing to be converted into industrial grade materials. Ferro-alloys such as Ferro-chromium and Ferro-silicon, for instance, are

produced by smelting mixed ores and are used in the production of various steels. Yet, one would be mistaken to think that alloying metals is a simple process. For example, if one were to simply mix molten aluminum with molten lead, they'd find that the two would separate into layers, much like oil and water.

Commercial and trade alloys generally require greater processing and are most often formed by mixing molten metals in a controlled environment. The procedure for combining molten metals or mixing metals with nonmetals varies greatly depending on the properties of the elements being used.

Because metal elements possess great variances in their tolerance of heat and gasses, factors such as melting temperatures of component metals, impurity levels, mixing environment and alloying procedure are central considerations for a successful alloy process.

While elements like the refractory metals are stable at high temperatures, others begin to interact with their environment, which can affect purity levels and, ultimately, the alloy quality. Often in such cases, intermediate alloys must be prepared in order to persuade elements to combine.

As an example, an alloy of 95.5 percent aluminum and 4.5 percent copper is made by first preparing a 50% mixture of the two elements. This mixture has a lower melting point than either pure aluminum or pure copper and acts as a "hardener alloy." This is then introduced to molten aluminum at a rate that creates the right alloy mix.

Uses of Alloys

Over 90% of metal use is in the form of alloys. Alloys are used because their chemical and physical properties are superior for an application than that of the pure element components. Typical improvements include corrosion resistance, improved wear, special electrical or magnetic properties, and heat resistance. Other times, alloys are used because they retain the key properties of component metals, yet are less expensive.

For example:

Steel: Steel is the name given to an alloy of iron with carbon, usually with other elements, such as nickel and cobalt. The other elements add a desired quality to the steel, such as hardness or tensile strength.

Stainless Steel: Stainless steel is another iron alloy, which typically contains chromium, nickel, and other elements to resist rust or corrosion.

18k Gold: 18 karat gold is 75% gold. The other elements typically include copper, nickel, and/or zinc. This alloy retains the color and luster of pure gold, yet is harder and stronger, making it better suited for jewelry.

Pewter: Pewter is an alloy of tin, with other elements such as copper, lead, or antimony. The alloy is malleable, yet stronger than pure tin, plus it resists the phase change of tin that can make it crumble at low temperatures.

Brass: Brass is a mixture of copper with zinc and sometimes other elements. Brass is hard and durable, making it suitable for plumbing fixtures and machined parts.

Sterling Silver: Sterling silver is 92.5% silver with copper and other metals. Alloying silver makes it harder and more durable, although the copper tends to lead to a greenish-black oxidation (tarnish).

Electrum: Some alloys, like electrum, occur naturally. This alloy of silver and gold was highly prized by ancient man.

Meteoritic Iron: While meteorites may consist of any number of materials, some are natural alloys of iron and nickel, with extraterrestrial origins. These alloys were used by ancient cultures to make weapons and tools.

Amalgams: Amalgams are mercury alloys. The mercury makes the alloy much like a paste. Amalgams may be used in dental fillings, with the mercury intact, although another use is to spread the amalgam and then heat it to vaporize the mercury, leaving a coating of another metal.

Substitutional Alloys

Substitutional alloys are metal alloys formed by substituting one metal atom for another metal atom of similar size. To form an alloy, the two substances (metals) should be mixed. The type and size of the substances that are mixed may determine the type of alloy that is formed. In order to form a substitutional alloy, the atoms of the two metals should be of similar size.

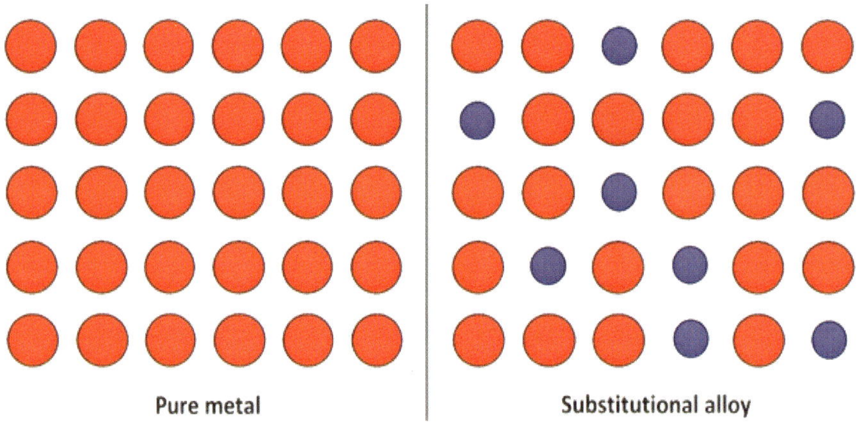

Figure: Comparison Between Pure Metal and Substitutional Alloy

A substitutional alloy is formed via atom exchange mechanism. Here, some metal atoms of a metal lattice are replaced or substituted by metal atoms of the other metal. But both types of metal atoms should be in similar or nearly similar sizes for this substitution to occur. The percentage of the difference between atoms should not exceed 15%.

A well-known alloy that is a substitutional alloy is Bronze. Bronze is an alloy that mainly consists of copper and tin metal. But sometimes, other elements are also mixed with copper to make bronze. Such elements include arsenic, phosphorous, aluminium, manganese and silicon. Copper and tin metal atoms are nearly similar in their sizes.

Why Substitutional Alloys Occur: Bonding

The bonding between two metals is best described as a combination of metallic electron "sharing" and covalent bonding, one can't occur without the other and the proportion of one to the other changes depending on the constituents involved. Metals share there electrons throughout there structure, this flow of electrons is the reason behind many of the characteristics associated with metals, including their ability to act as conductors. The different amount and strength of covalent bonds can change depending on the different specific metals involved and how they are mixed. The covalent bonding is what is responsible for the crystal structure as well as the melting point and various other physical properties.

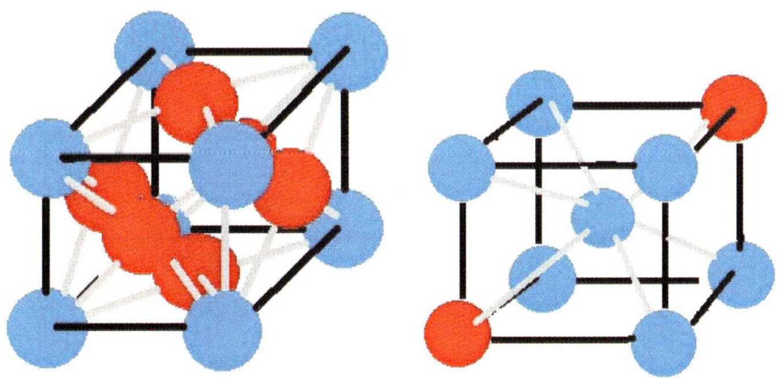

Figure: Examples of Substitutional Metal Alloys.

Depending on the specific type of substitutional alloy they can have multiple crystal structures. Two of the possible structures include Face Center Cubic (left), and Cubic Center Cubic (right). The structure of the metal alloy is not limited to these two structures, but combined they represent a large portion of the common alloys.

As the similarities between the electron structures of the metals involved in the alloy increase, the metallic characteristics of the alloy decrease. Pure metals are useful but their applications are often limited to each individual metal's properties. Alloys allow metal mixtures that have increased resistance to oxidation, increased strength, conductivity, and melting point; essentially any property can be manipulated by adjusting alloy concentrations. An example could be Brass Door fixtures they are strong and resist corrosion better then pure zinc or copper, the two major metals that constitute a brass alloy. The combination also has a low melting point allowing it to be easily cast into many different shapes and sizes. There are many other aspects of substitutional alloys that could be explored in depth, but the basic concept is the idea that each individual metal in an alloy give the final product its chemical and physical properties.

Substitutional alloys played an important role in the development of human society and culture as we know it today. The Bronze age itself is named after the Substitutional alloy consisting of tin in a metallic solution of copper. Ancient bronzes are very impure, or even mislabeled, containing large amounts of zinc and arsenic as well as lots of impurities. These many substitutional alloys allowed for stronger tools and weapons, they allowed for increased productivity in the workshop as well as on the battlefield. The need for raw materials like tin and copper for the production of bronze also spurred an increase in trade, since their ores are rarely found together. The current chemical understanding of substitutional alloys would not be so in depth if it weren't for the usefulness of the alloys to humans.

Interstitial Alloys

Interstitial alloys are metal alloys formed by inserting smaller atoms into holes of the metal lattice. A metal lattice is a metallic structure that consists of a giant lattice of positively charged ions and delocalized outer electrons. An interstitial alloy is formed when a metal is mixed with another substance composed of small atoms (having small radii) that can reside in the holes of the metal lattice.

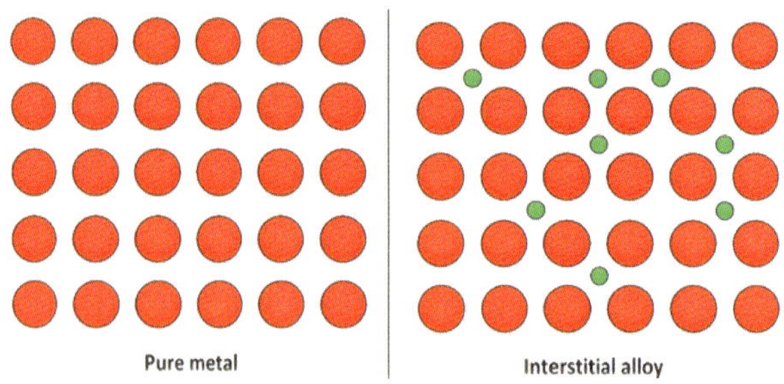

Figure: Comparison of a Pure Metal with an Interstitial Alloy

Some examples of such small atoms include hydrogen, carbon, boron, and nitrogen. The holes in a metal lattice are the spaces between the metal atoms. Steel is a good example of an interstitial alloy. Steel is a metal alloy containing iron, carbon and some other elements. Here, the small carbon atoms fill the spaces between large iron atoms. Because of this tightly packed structure of steel, it is hard and very strong.

The formation of an interstitial alloy is via the interstitial mechanism. Here, one type of atom is considerably smaller than the other type. Since these atoms are much smaller, they cannot replace atoms of the metal lattice as for substitutional alloys. Instead, they get trapped in spaces the large atoms of the metal lattice. These spaces are called interstices.

Amalgam

Mercury alloyed with a number of metals forms amalgam, which is primarily used for dental fillings. The common constituents of amalgam are mercury, tin, silver, copper and other trace metals. The use of amalgam in the dental sector dates back to the 1800s since it is easy to apply, has good durability and strength and is economical.

The popularity of amalgam has diminished in recent years especially due to its Mercury content and since reliable composite materials have become widely available.

Chemical Properties

The composition of amalgam is given in the table below:

Silver	40-60%
Tin	27-30%
Copper	13-30%
Zinc	1%

High-copper amalgam includes 40-60% silver, 27-30% tin and 13-30% copper and 1% zinc set with mercury. Indium and palladium are also included. Silver enables setting expansion and causes an increase in strength and resistance to corrosion. Tin may cause setting contraction whereas copper improves strength, minimizes corrosion and tarnish, brings down creep and brings down cases of marginal leakage. Zinc brings down the oxidation of other alloys in the metal. It has been proved that zinc-containing amalgams have a longer life when compared to non-zinc amalgams. Indium reduces creep and increases strength whereas palladium reduces corrosion and tarnish.

Tarnishing is the loss of luster from a metal or alloy surface because of the formation of a surface coating. The alloy remains unchanged and the mechanical properties also remain the same. A sulphide layer formed in the surface causes amalgam to tarnish.

Galvanic corrosion takes place when two dissimilar metals are present in a wet environment. The flow of electric current between the metals results in one of the metals getting corroded. Surface corrosion may cause a change of color of an amalgam restoration and may result in pitting. Surface corrosion fills the amalgam/tooth interface with corrosion products bringing down micro leakage. Internal corrosion will result in marginal breakdown and fracture. Galvanic corrosion is promoted by an acidic environment.

Mechanical Properties

An amalgam restoration needs to be strong enough so as to resist the biting forces of occlusion. Dental amalgam has high compressive strength, which is 380MPa for low-copper amalgam and 414MPa for high-copper amalgams. However shear and tensile strengths are quite low. Hence it in essential that tooth structures support the amalgam to ensure long-term clinical success.

Creep is a gradual change in shape due to compression because of dynamic intra-oral stresses. Creep causes the flow of amalgam, hence unsupported amalgam will protrude from the cavity margin. These unsupported edges are quite weak and corrosion may cause further lowering of strength. Creep may also result in an overhang on fillings causing food to be trapped leading to secondary decay. The gamma-2 phase of amalgam is key to the considerably high creep values exhibited by certain materials.

Dimensional change is the net expansion or contraction of an amalgam. In case the amalgam contracts, dimensional change is negative and is positive in case of expansion during setting. According to the ADA/ANSI specification, the dimensional change should be a contraction or expansion of not more than 20 μm/cm.

Dimensional change is impacted by a number of factors such as the alloy/mercury ratio and also condensation and trituration techniques. Contraction and expansion take place simultaneously during the amalgamation reaction.

Types of Amalgam

Zinc Amalgam

Zinc amalgam finds use in organic synthesis (e.g., for the Clemmensen reduction). It is the reducing agent in the Jones reductor, used in analytical chemistry. Formerly the zinc plates of dry batteries were amalgamated with a little mercury to prevent deterioration in storage.

Potassium Amalgam

For the alkali metals, amalgamation is exothermic, and distinct chemical forms can be identified, such as KHg and KHg_2. KHg is a gold-coloured compound with a melting point of 178 °C, and KHg_2 a silver-coloured compound with a melting point of 278 °C. These amalgams are very sensitive to air and water, but can be worked with under dry nitrogen. The Hg-Hg distance is around 300 picometres, Hg-K around 358 pm.

Phases K_5Hg_7 and KHg_{11} are also known; rubidium, strontium and barium undecamercurides are known and isostructural. Sodium amalgam ($NaHg_2$) has a different structure, with the mercury atoms forming hexagonal layers, and the sodium atoms a linear chain which fits into the holes in the hexagonal layers, but the potassium atom is too large for this structure to work in KHg_2.

Sodium Amalgam

Sodium amalgam is produced as a byproduct of the chloralkali process and used as an important reducing agent in organic and inorganic chemistry. With water, it decomposes into concentrated sodium hydroxide solution, hydrogen and mercury, which can then return to the chloralkali process anew. If absolutely water-free alcohol is used instead of water, an alkoxide of sodium is produced instead of the alkali solution.

Aluminium Amalgam

Aluminium can form an amalgam through a reaction with mercury. Aluminium amalgam may be prepared by either grinding aluminium pellets or wire in mercury, or by allowing aluminium wire or foil to react with a solution of mercuric chloride. This amalgam is used as a reagent to reduce compounds, such as the reduction of imines to amines. The aluminium is the ultimate electron donor, and the mercury serves to mediate the electron transfer. The reaction itself and the waste from it contain mercury, so special safety precautions and disposal methods are needed. As an environmentally friendlier alternative, hydrides or other reducing agents can often be used to accomplish the same synthetic result. Another environmentally friendly alternative is an alloy of aluminium and gallium which similarly renders the aluminium more reactive by preventing it from forming an oxide layer.

Tin Amalgam

Tin amalgam was used in the middle of the 19th century as a reflective mirror coating.

Other Amalgams

A variety of amalgams are known that are of interest mainly in the research context.

- Ammonium amalgam is a grey, soft, spongy mass discovered in 1808 by Humphry Davy and Jöns Jakob Berzelius. It decomposes readily at room temperature or in contact with water or alcohol:

$$2\ H_3N\text{-}Hg\text{-}H \xrightarrow{\Delta T} 2\ NH_3 + H_2 + 2\ Hg$$

- Thallium amalgam has a freezing point of −58 °C, which is lower than that of pure mercury (−38.8 °C) so it has found a use in low temperature thermometers.
- Gold amalgam: Refined gold, when finely ground and brought into contact with mercury where the surfaces of both metals are clean, amalgamates readily and quickly to form alloys ranging from $AuHg_2$ to Au_8Hg.

Dental Amalgam

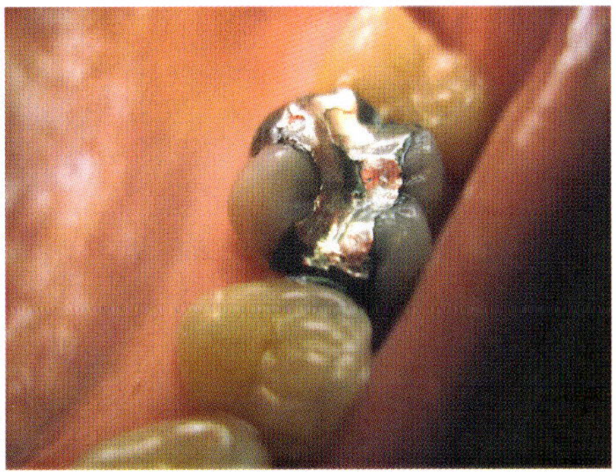

An amalgam dental filling

Dentistry has used alloys of mercury with metals such as silver, copper, indium, tin and zinc. Amalgam is an "excellent and versatile restorative material" and is used in dentistry for a number of reasons. It is inexpensive and relatively easy to use and manipulate during placement; it remains soft for a short time so it can be packed to fill any irregular volume, and then forms a hard compound. Amalgam possesses greater longevity when compared to other direct restorative materials, such as composite. However, this difference has decreased with continual development of composite resins.

Amalgam is typically compared to resin-based composites because many applications are similar and many physical properties and costs are comparable.

Manufacturing Process

For the preparation of dental amalgams, liquid mercury is mixed with a powdered alloy mainly comprising tin and silver. Lathe cutting or milling of a cast ingot of the silver-tin alloy is done to produce the powder. The particles formed are shaped irregularly.

Alternatively, the liquid alloy may be atomized and condensed, a process that causes particles with a spherical morphology. Both these alloy types are used in clinical amalgams as a mixture of spherical and lathe-cut particles. For clinical purposes, amalgam is mixed with mercury in a process called trituration. Previously, they were formed manually; however, presently vibratory mixers are present and manufacturers prepare the unmixed amalgam in two chambers of a tiny capsule.

Before mixing, the thin membrane separating the alloy powder from the liquid mercury is destroyed and the capsule is placed in the mechanical mixer arm and vibrated for a specific amount of time to enable thorough mixing of liquid and powder. The extrusion of the freshly mixed amalgam with a plastic consistency is done from the capsule into the cavity. During the trituration process, the surface layer of the silver-tin alloy dissolves in the liquid mercury and the reaction causes new phases to be formed. These new solid phases cause solidification of the plastic amalgam paste.

Applications

The applications of amalgams are the following:

- Dental amalgam is an excellent and versatile restorative material and is used in dentistry for several reasons. It is economical and quite easy to use and can be manipulated during placement. Amalgam remains soft for a brief time period so it is possible to fill it into any irregular volume and then a hard compound is formed. Amalgam has a greater longevity when compared to other direct restorative materials, like composites.
- Sodium amalgam has been used in organic chemistry as a powerful reducing agent, which is much safer than sodium. Sodium amalgam is also used while designing the high pressure sodium lamp providing sodium to obtain the right color and mercury to fine-tune the electrical features of the lamp.
- Aluminium amalgam is used as a reducing agent.
- Thallium amalgam finds use in low-temperature thermometers.

Use in Mining

Mercury has been used in gold and silver mining because of the convenience and the ease with which mercury and the precious metals will amalgamate. In gold placer mining, in which minute specks of gold are washed from sand or gravel deposits, mercury was often used to separate the gold from other heavy minerals.

After all of the practical metal had been taken out from the ore, the mercury was dispensed down a long copper trough, which formed a thin coating of mercury on the exterior. The waste ore was then transferred down the trough, and gold in the waste amalgamated with the mercury. This coating would then be scraped off and refined by evaporation to get rid of the mercury, leaving behind somewhat high-purity gold.

Mercury amalgamation was first used on silver ores with the development of the patio process in Mexico in 1557. There were also additional amalgamation processes that were created for processing silver ores, including pan amalgamation and the Washoe process.

Gold Amalgam

Gold Extraction

Gold amalgam has proved effective where gold fines ("flour gold") would not be extractable from ore using hydro-mechanical methods. Large amounts of mercury were used in *placer mining*, where deposits composed largely of decomposed granite slurry were separated in long runs of "riffle boxes", with mercury dumped in at the head of the run. The amalgam formed is a heavy solid mass of dull gray color. (The use of mercury in 19th century placer mining in California, now prohibited, has caused extensive pollution problems in riverine and estuarine environments, ongoing to this day.) Sometimes substantial slugs of amalgam are found in downstream river and creek bottoms by amateur wet-suited miners seeking gold nuggets with the aid of an engine-powered water vacuum mounted on a float.

Gold Extraction

Interior of the Deadwood Terra Gold Stamp Mill.

Crushed ore is washed over mercury coated copper sheets, and fine gold particles form an amalgam with the mercury. The amalgam was scraped off and the gold then separated from the amalgam by heating and evaporating the mercury which was then recovered by a condenser for reapplication to the plates

Where stamp mills were used to crush gold-bearing ore to fines, a part of the extraction process involved the use of mercury-wetted copper plates, over which the crushed fines were washed. A periodic scraping and re-mercurizing of the plate resulted in amalgam for further processing.

Gold Extraction

Amalgam obtained by either process was then heated in a distillation retort, recovering the mercury for reuse and leaving behind the gold. As this released mercury vapors to the atmosphere, the process could induce adverse health effects and long term pollution.

Today, mercury amalgamation has been replaced by other methods to recuperate gold and silver from ore in developed nations. Hazards of mercurial toxic waste have played a major role in the phasing out of the mercury amalgamation processes. However mercury amalgamation is still regularly used by small-scale gold placer miners (often illegally) & particularly in developing countries.

Amalgam Probe

Mercury salts are, compared to mercury metal and amalgams, highly toxic due to their solubility in water. The presence of these salts in water can be detected with a probe that uses the readiness of mercury ions to form an amalgam with copper. A nitric acid solution of salts under investigation is applied to a piece of copper foil, and any mercury ions present will leave spots of silvery-coloured amalgam. Silver ions leave similar spots but are easily washed away, making this a means of distinguishing silver from mercury.

The redox reaction involved where mercury oxidizes the copper is:

$$Hg^{2+} + Cu \rightarrow Hg + Cu^{2+}.$$

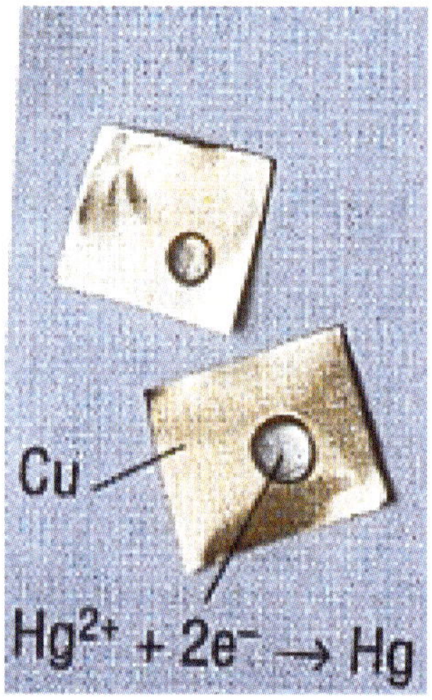

The amalgam probe

Amorphous Metal

Amorphous metal alloys get their name from their amorphous atomic structure. The material has no crystal structure as a result. The amorphous atomic structure is characteristic of glass glasses, so the material may be referred to as metal glass. The amorphous structure refers to the disordered arrangement of atoms in the metal. They generally exhibit greater hardness, Yield (engineering) |yield and fracture stresses as well as a comparable modulus of Elastic modulus elasticity and Shear modulus |rigidity than crystalline metals. The strength comes from the lack of Slip (materials science) |slip planeson Grain boundary grain boundaries that are as a result of a Crystal structure crystal structure.

The amorphous metal structure was discovered in the 1950's when lead foil was Quench quenched at roughly 10^{12} K/s (degrees Kelvin per second). This resulted in avoiding crystallization during solidification, preserving the amorphous arrangement of the atoms that existed in the metal's liquid state. The cooling of a part requires the dissipation of heat to its surroundings, so the geometry of the part would largely affect its ability to be quenched at a quick rate. The formation of the amorphous structure was limited to foils approximately 50 µm thick.

However, the unique properties of the metal achieved by this process would warrant the creation of a new class of materials: metal glasses, due mainly to their amorphous atomic structure as is characteristic of glasses. The formation of bulk metallic glasses (BMGs) of several centimeters thick soon occurred with advances in processes and materials. In 1969, Chen and Turnbull formed amorphous spheres of $Pd_{77.5}Cu_6Si_{16.5}$ at critical cooling rates of 100 - 1000 °K/s with diameters of 0.5mm.

Forming

The major principle contributing to the formation of metallic glass is the quick cooling rate. Longer cooling rates from the Melting point melting point will result in larger grains because the atoms have more time to order themselves into crystals. The rate of cooling can increase to the point that the grains would not only become very small, but soon cease to exist all together. The atom's amorphous arrangement that existed in the metal's liquid state is preserved.

Since the key to the formation of the amorphous structure seems to be the prevention of the formation of crystals, this objective can be addressed directly through disrupting the forces that form the Metallic bonds themselves. Metals of vastly different size will have difficulty bonding, so an alloy containing them would take longer to crystallize. This arrangement of a vast variety of atomic radii in the same alloy is referred to as the "confusion principle". For the glass to form, the atomic radii of the different metals must differ by at least 12%. Alloys containing an array of elements such as Zirconium, Aluminum, Nickel and Copper can reach critical cooling rates of as low as 1 K/s. The critical cooling rate refers to the lowest rate of cooling at which an amorphous structure can be attained.

An additional strategy to encourage the formation of glass is choosing the composition of the alloy based on the melting temperature (T_m) and Glass transition glass formation temperature (T_g). The lower T_m and higher T_g is for the alloy, the smaller the difference between the two temperatures, and therefore, the less time required to form a glass at a potentially lower cooling rate. Related temperatures for the formation of BMGs are outlined in table below:

Thermal properties and glass forming ability of Fe Cr Mo P C B BMGs

Alloy	D (mm)	T_g (K)	T_x (K)	ΔT_x (K)	T_m (K)	T_l (K)	T_{rg}
$Fe_{71}Mo_5P_{12}C_{10}B_2$	3	701	733	32	1199	1257	0.558
$Fe_{69}Cr_2Mo_5P_{12}C_{10}B_2$	3	710	744	34	1210	1266	0.561
$Fe_{69}Mo_7P_{12}C_{10}B_2$	2.5	716	753	37	1202	1263	0.567
$Fe_{66}Mo_{10}P_{12}C_{10}B_2$	2.5	721	757	36	1200	1267	0.569
$Fe_{64}Cr_3Mo_{10}P_{10}C_{10}B_3$	2.5	729	775	46	1228	1288	0.566
$Fe_{63}Cr_3Mo_{10}P_{12}C_{10}B_2$	2.5	735	778	43	1219	1277	0.576
$Fe_{63}Cr_3Mo_{12}P_{10}C_7B_5$	2.5	747	791	44	1230	1299	0.575
$Fe_{65}Cr_2Mo_9P_{10}C_8B_6$	2.5	756	820	64	1232	1288	0.587

T_g and T_x are determined using DSC. T_m and T_l are measured by DTA. Glass forming ability is represented by the maximum attainable diameters (D) of glassy rods.

Melt spinning, one of the first process for producing a metal glass, achieved cooling rates on the order of >1000 °K/s. This was necessary for alloys that required such rates for glass formation to occur. The process consisted of pouring a stream of the molten metal over a rapidly rotating drum. The drum was cooled internally with liquid nitrogen. The quick rotation of the drum enabled a thin application of the metal on the drum and for a short amount of time. It was possible to cool a thin quantity of metal that quickly with a conductive method of cooling.

More modern alloys do not require such high cooling rates. This property makes casting a possibility because sufficient cooling can be provided by the walls of moulds. Liquid nitrogen can be used as a powerful agent for cooling the mould wall. Heat can only be dissipated by the outer surface of the metal, so the cooling is still highly dependent on the geometry of the metal part. As shown in figure, the formation of the amorphous state is the ability for the cooling curve (the blue line for example) to "miss" the crystalline nose. Lowering the melting temperature (T_1 as illustrated in the plot) and raising or maitaining the glass formation temperature, as shown in figure, would yield a cooling curve with a lesser slope, and therefore lower cooling rate, that misses the cryslalline nose.

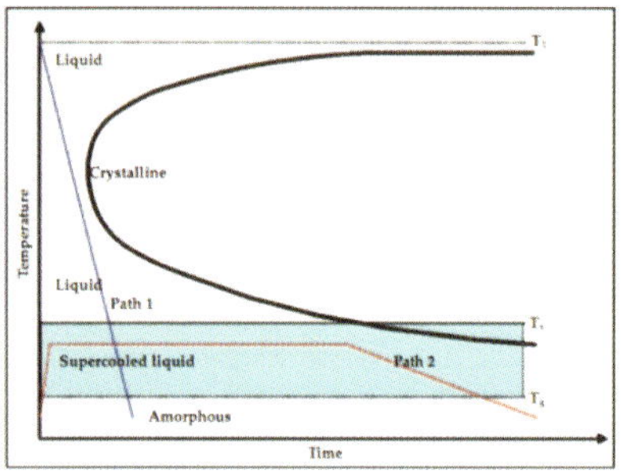

Figure: Cooling diagram.

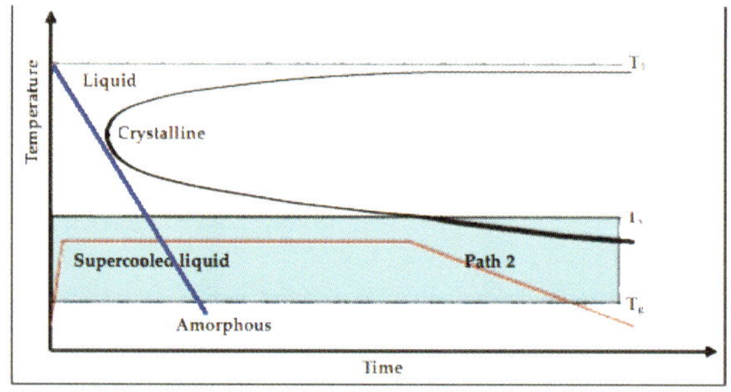

Figure: Cooling diagram showing lower melting temperature.

Mechanical Properties

The amorphous structure of these glasses results in a lack of Slip (materials science) slip planes that would normally exist in a material with Crystallite grains and Grain boundary grain boundaries. As a result, metal glasses exhibit much greater resistance to Deformation (mechanics) deformation than crystalline metals. This property generally leads to a much greater Vickers hardness test Vickers hardness (H_v), Yield (engineering)|yield stress (σ_Y) and Fracture stress (σ_f) as demonstrated by table. Other properties outlined are the modulus of Elastic modulus elasticity (E), modulus of Shear modulus rigidity (G), Bulk modulus bulk modulus (K) and Poisson's ratio poisson's ratio (v) which are comparable to that of other existing engineering materials.

Mechanical properties of Fe Cr Mo P C B BMGs

Alloy	H_v	σ_y (GPa)	σ_f (GPa)	ε_{pl} (%)	G (GPa)	K (GPa)	E (GPa)	ν
$Fe_{71}Mo_5P_{12}C_{10}B_2$	845	2.45	3.15	3.6	62			
$Fe_{69}Cr_2Mo_5P_{12}C_{10}B_2$	864	2.6	3.2	2.2	66			
$Fe_{69}Mo_7P_{12}C_{10}B_2$	882	2.6	3.25	2.8	65			
$Fe_{66}Mo_{10}P_{12}C_{10}B_2$	900	2.55	3.25	1.7	66.1	176	176	0.333
$Fe_{64}Cr_3Mo_{10}P_{10}C_{10}B_3$	950	2.75	3.4	1.2	66	174	176	0.33
$Fe_{63}Cr_3Mo_{10}P_{12}C_{10}B_2$	931	2.6	3.4	3.2	65.3	178	178	0.34
$Fe_{63}Cr_3Mo_{12}P_{10}C_7B_5$	974	2.9	3.5	0.9	68.8	179	183	0.33
$Fe_{65}Cr_2Mo_9P_{10}C_8B_6$	973	2.9	3.55	1.0	66.8	174	177	0.33

Symbols for stress strain curve parameters obtained from compression tests: σ_y = yield stress, σ_f = fracture strength, and ε_{pl} = plastic strain. The measurement errors in G, K, E, and ν are estimated to be in the range $\pm$(0.6 0.7) GPa , $\pm$(3 4) GPa, $\pm$(4 5) GPa , and $\pm$(0.01 0.012), respectively.

Table: Stress-strain curve parameters obtained from compression tests of BMGs.

Grain boundary: Grain boundaries are also a weak point for Corrosion as they provide more surface area for the required chemical reactions to occur. The lack of grain boundaries in BMGs reduce their tendency to corrode.

For comparison, properties of some common engineering materials are given in Table below.

Material		σ_Y (MPa)	G (GPa)	E (GPa)	ν	ρ (Mg/m³)
Aluminum	6061-T6	255	26	68.9	0.35	2.71
Cast Iron	ASTM A-197	-	68	172	0.28	7.28
Copper	Bronze C86100	345	38	103	0.34	8.83
Magnesium	Am 1004-T61	152	18	44.7	0.3	1.83
Steel	Structural A36	250	75	200	0.32	7.85
	Stainless 304	207	75	193	0.27	7.86
	Tool L2	703	75	200	0.32	8.16
Titanium	Ti-6Al-4V	924	44	120	0.36	4.43

Table: Properties of common engineering materials.

It is evident that the Iron based BMGs examined in table have approximately 3 times the yield stress of titanium Ti-6Al-4V and over 10 times the yield stress of structural steel A36 with comparable modulus of elasticity and rigidity. Their high yield stress and therefore high resistance to deformation gives metal glasses very high elasticity and ability to store mechanical energy.

Because of the stiffness of BMGs, they are considered to have poor ductility and therefore poor tensile strength. When a significant tensile load is applied, they experience a phenomenon refered to as shear banding which occurs as a result of localized shear.

Forming Methods

Die Casting

When die casting materials such as iron, their net volume can reduce significantly over the cooling process, yielding an inaccurate part and therefore requiring surface finishing surface finishing afterwards. Die casting metal glass parts is very practical because there is almost no shrinkage. This happens for two main reasons. Firstly, the formation of glass is such that the arrangement of atoms in glass is the same as the liquid, so there is essentially no Phase transition phase change. Since the atoms do not rearrange, the volume of the die does not change and therefore no shrinkage. Secondly, the alloys possessing low T_m will require less cooling than most Carbon steel carbon steels for example. A smaller temperature change also leads to less contraction during cooling. These properties of metal glasses will lead to near net shape near net shape die cast parts with little need for additional Surface finishing surface finishing after casting.

Thermoplastic Forming

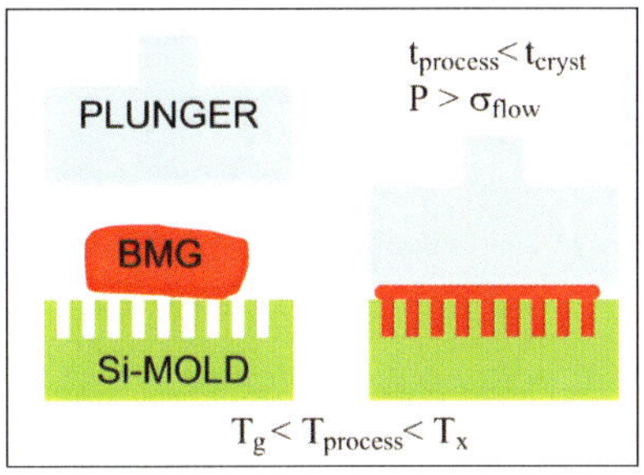

Figure: A rough schematic of the elements involved in the thermoplastic forming of BMGs.

BMGs can be formed into relatively complicated shapes by thermoplastic forming. When the glass is at a temperature slightly above T_g, it is plastic enough to be deformed without fracturing. An ingot of metallic glass is pressed into a die as shown in figure. Since the forming process occurs at above T_g, the glass has not set and can still crystallize if left at this temperature for a too long. Certain alloys of BMG's are resistant enough to crystallizing that enough time exist to carry out the thermoplastic forming process while maintaining desired the properties of the part. We know that BMGs are limited to smaller geometriess but they also have low shrinkage during cooling. These traits make thermoplastic forming applicable for the formation of small parts. The average pressure required to carry out the formation of a part is expressed by the Hagen–Poiseuille equation.

$$p = 16v.\eta.\frac{L}{d^2}$$

L is the length of a channel, d is the diameter, η is the viscosity of the BMG, v is the velocity at which it is moving down the channel and p is the pressure required for the process. The model suggests that the feasible maximum forming pressure and therefore minimum feature diameter are approximately 300 MPa and 10nm.

Process Energy Reduction

The most obvious method of reducing the energy required to produce amorphous metals is to create metals with lower melting points. If the melting point is lower, less energy has to go into heating the metal up to the appropriate temperature. Conversely, the energy efficiency of the process for rapidly cooling the metal would be improved by the lower melting temperature because less energy would be required to cool the metal over a lesser temperature difference. A slower critical cooling rate is more likely with a lower melting temperature so more of the effect of cooling can be lent to the ambient temperature. A substantial amount of energy is required to force cooling to achieve faster cooling rates than that which would be achieved by conventional quenching. Decreasing the critical cooling rate primarily would reduce energy consumption for the mentioned reasons.

The implementation of Squeeze casting to the process of producing BMGs can increase its efficiency. Squeeze casting is essentially performing a casting process at pressures on the order of 100Mpa. During solidification, high pressure causes the liquid metal to maintain contact with the mould wall throughout the process. The mould wall provides the cooling to the material. This improves the efficiency of heat removal, increasing the rate of cooling that can be achieved. The melting temperature and the glass forming temperature of metals increase at higher pressures. So melting the metal at a lower pressure and casting it at an increased pressure would essentially decrease the temperature difference that the cooling process would have to achieve.

Methods for increasing the efficiency of the thermoplastic forming process can be hypothesized upon inspection of the Hagen-Poiseuille equation. An increase in efficiency would be achieved it the velocity of the process were increased and the pressured required were decreased or maintained. A higher velocity would decrease the amount of time that the thermoplastic forming process requires, lending more time to the cooling process once the final shape of the material has been established. This could easily be obtained by changing the geometry of the part that is to be formed. Decreasing the length and increasing its diameter while maintaining the pressure would increase the velocity of the forming process. Decreasing the amount and intricacy of features on the part will yield the same effect as increasing the diameter. Decreasing the viscosity of the metal during the process would increase its efficiency. This could be achieved by either changing its composition to less viscous metal or by performing the process at as close to the glass forming temperature as possible.

Limitations

Despite several positive traits of metallic glasses, several limitations of the material still exist. Due to its high strength, it generally exhibits a low elastic strain limit. The material will not deform under load, but rather fail catastrophically when the fracture stress is reached. This can be dangerous in structural application because few visual cues are given if the material is about to fail.

Though the formation of metallic glasses has evolved from thin impractical foils to modern BMGs of several centimeters in diameter, they are still limited to parts of small thickness and awkward geometry.

Various process for producing metallic glasses are still in their infancy, so adoption and therefore production of the material is not yet wide spread. Production costs are far higher than conventional crystalline alloys so they are limited to applications. The process can become more economically feasable in the future as facilities are expanded to accommodate the up scaling of the process. Just like today's mass produced materials such as steel, producing BMGs in such high quantities would likely offset much of the overhead costs per unit of product associated with small scale production.

Metal glasses are limited to low temperature applications because of their characteristically low glass formation temperatures. If they were to be placed in an environment that exceeds their glass formation temperature, they would loose their amorphous property and likely reform into a crystalline metal.

Applications

Metal glasses have found applications in high end markets where any performance gain from utilizing them in the product is justified, regardless of the cost. They have begun to appear in higher end electronics as cases due to their rigidity, hardness and therefore scratch resistance. The high hardness is ideal for use in tooling. The lack of grain structure allows a blade to be sharpened to an exceptional edge because there is no length scale above the atomic to limit it. This property is useful in knives, especially in scalpels.

High elastic energy storage per unit volume and mass, and the low damping, give metallic glasses potential as springs. Sporting equipment such as golf clubs and baseball bats utilize the high hardness and elastic energy property for good energy transfer to projectiles.They was successful used in golf club heads and in tennis racket frames where the property is exploited. Other possible applications of springs in devices are high-speed relays.

Information storage and reproduction would utilize the lack of grain structure and high hardness. Features of near-atomic scale could be moulded or etched into a metallic glass surface to make masters for reproducing ultra-high density digital data.

Small parts can be die-cast to achieve a near net-shape fabrication where additional machining to achieve the shape would be expensive and impractical.

Oxide Dispersion Strengthened Alloy

Oxide dispersion strengthened alloys (ODS alloys) typically consist of a high temperature metal matrix - such as iron aluminide, iron chromium, iron-chromium-aluminium, nickel chromium or nickel aluminide - with small (5-50nm) oxide particles of alumina (Al_2O_3) or yttria (Y_2O_3) dispersed within it.

Iron-based and nickel-based oxide dispersion strengthened alloys exhibit good corrosion resistance and mechanical properties at elevated temperatures. These alloys also show excellent creep resistance, which stems partly from the dispersion of oxide and other particles, and partly from the very large elongated grain structure.

Properties of Oxide Dispersion Strengthened (ODS) Alloy

The compositions and physical properties of two commercially-available oxide dispersion strengthened (ODS) alloys (one iron-based, the other nickel-based) are given in the tables below.

	Composition (wt%)						
Name	**Ni**	**Fe**	**Cr**	**Al**	**Ti**	**C**	**Yttrium Oxide**
INCONEL MA754	78	1.0	20	0.3	0.5	0.05	0.6
INCOLOY MA956	-	74	20	4.5	0.5	0.05	0.5

The physical properties of the above materials, at room temperature and at 1000°C, are shown in the next table:

Property	**MA754**		**MA956**	
Melting point (°C)	1400		1482	
Density (gcm^{-3})	8.3		7.2	
	Room Temp	1000°C	Room Temp	1000°C
Coefficient of thermal expansion (/K)	$12.2x10^{-6}$	$17.0x10^{-6}$	$11.3x10^{-6}$	$14.9x10^{-6}$
Thermal conductivity (W/m.K)	14.3	32.6	10.9	25.5
Specific heat (J/kg.K)	440	695	469	741
Electrical resistivity (μ Ω.m)	1.08	1.16	1.31	1.43
Modulus of elasticity (GPa)	203	133	269	191

ODS alloys are specifically designed, through a combination of composition and microstructure, to operate at high temperature under mechanical stress for extended periods. This requires a combination of excellent chemical resistance, particularly to high temperature oxidation, with retention of mechanical properties at high temperature, a function of creep resistance and elastic modulus achieved through elongated grain structure and grain boundary pinning.

Production of Oxide Dispersion Strengthened (ODS) Alloy

Oxide dispersion strengthened (ODS) alloys are produced by the mechanical alloying of powders. The powder constituents can be in the elemental, intermetallic or pre-alloyed state.

The metal and oxide (i.e. Yttria) powders are blended and mechanically alloyed using a ball mill. The repeated impacts of this process cause smearing together of the powders followed by shearing which generates particles that each contain a fine mixture of the constituents.

The powder is then packaged into sealed containers that are hot worked into simple primary shapes, either by extrusion or hot isostatic pressing (HIP). The resultant product is dense and fine grained (less than 1μm) but has highly developed, directional residual strain. The alloys are then re-crystallized at about 1280-1350°C when large and highly elongated grains form.

Shape-memory Alloy

Shape memory alloys are the designed materials in which large distortion can be stimulated and regained through temperature changes or stress changes which results martensitic phase change and induced elasticity at high temperatures. These special materials kept on being developed for intensify the accomplishment as per the need for the engineering field. Shape-memory alloys (SMAs) are a class of novel materials that exhibit two outstanding unique properties namely the shape-memory effect and super elasticity. These properties make them different from the ordinary materials based on the diffusion less transformation in solid. Shape memory effect is one of the unique property which returns to the actual shape during heating, after being deformed, In other words it has the ability of the material to regain from plastic deformation through thermal processing. During hysteretic loop, super elasticity has the ability to recover from large elastic strains (up to 8%) also permits the material to resist large cyclic deformations, without residual strains. due to these immanent wonderful properties, SMAs are adopted in new technological applications related to energy dissipation in civil engineering structural designs, vibration control devices.Super elasticity is obtained in a limited temperature range just above its transformation temperature, heating is not required for getting un deformed shape to recover and exhibits enormous elasticity, associated with large non-linear recoverable strains. The technology pushes towards 'Smart materials 'with adaptive or innovative functions makes use as sensors, actuators and micro-controllers, which results a disagreeable increase in weight and volume of the related machine components. The development of high 'functional density' and 'smart materials' applications must conquer technical and commercial provisions, such as available space, environmental conditions, response time and allowable cost.

Methods of Preparation of Shape Memory Alloys

Melting

SMA can be produced by many ways one of the most important method is vacuum melting. In addition to the vacuum melting other melting techniques like electron beam melting and plasma melting can also be adopted. Vacuum induction melting works on Electromagnetic induction works by inducing electrical eddy currents in the graphite crucible and in the metallic charges. Furthermore electro dynamic force is adopted for excellent stirring and mixing of the melt. NiTi alloys are melted by VIM in graphite crucible carbon content increases and forms TiC in the alloy, due to this transformation temperature decreases. By using electro graphite crucible SMA can achieve low carbon concentrations. The advantage of VIM is achieving of chemical composition homogeneity throughout the ingot and controllability of chemical composition. Vacuum arc re-melting (VAR) doesn't have problem of contamination of carbon in the alloy due to the absence of graphite crucible and gives high purity. VAR classified in to two types with respect to the heating system first method is non consumable electrode and is preferred in laboratories and is applicable to many varieties of alloys in this method raw materials to be melted are placed in a copper mould and irradiated by the argon arc from an electrode made of a tungsten rod. When the alloy melts it resembles button shape due to the surface tension effect, the button shape again re-melted number of times in order to get homogeneity composition. Second method is consumable electrode consists of two sources one is heating source and another one is material source the electrode is heated by the argon arc and the molten alloy drops down on to the mould and forms a cylindrical ingot. A drawback of melting multiple types results oxygen and carbon pick up if any leakage of vacuum occurs. Double melting process using VIM primary followed by VAR is often used to get further refining. Electron beam melting utilizes high voltage electron beam as a heating source this method is mainly adopted for NiTi SMA which doesn't need accurate or definite control of transformation temperature. Plasma melting method uses a low viscosity electron beam discharged from a hollow plasma cathode, this method results very less loss of the alloy element .composition distribution is uniform inspite of the use of water cooled copper mould.

Figure: Arc Melting Unit

Heat Treating

NiTi samples must be cold worked before heat treatment in order to achieve optimized properties. If we consider super elastic nitinol, heat treatment may be adopted at a temperature of 500°C, for shape memory alloys it usually lies between 350°C- 450°C. Unique properties like SME and pseudo elasticity is good at a Nickel composition of 55.5(wt %). At high temperatures these properties will achieved by solution treatment and aged at a temperature around 400°C rich Ni- phases precipitates and as the composition of matrix varies transformation temperature increases.

Powder Metallurgy

Powder metallurgy is an essential technique for producing near net shape components of NiTi alloys. In case of copper based shape memory alloys a grain refinement to enhance fatigue properties is targeted. There are two kinds of powder metallurgy techniques for preparing NiTi alloys, one is raw metal powder sintering and another one is alloy powder sintering. The homogeneity of NiTi alloy can be improved compared with raw metal powder sintering. Different powder metallurgy methods have been developed for Ni Ti in which both elemental powders and pre alloyed powders can be used as starting materials by hot isostatic pressing (HIP), hydriding and pulverization. HIP is most commonly adopted sintering technique powders are filled in evacuated and gas tight welded cans will transfer the pressure to the powder. Samples prepared by this method having combined advantages of theoretical density and high homogeneity. If we require high amount of complex near net shape parts Metal injection moulding (MIM) becomes a cost effective. It is a combination of polymer injection moulding and powder metallurgy (PM).due to low density their yield strain is only 1%.

Thermal Spray

Thermal spray process is mainly used for producing NiTi foils or NiTi tubes, thin walled mill products and other 3D shapes. It offers near net shape processing to reach the advantage of processing of highly reactive materials under controlled conditions. This process can be done by two methods one is low pressure wire arc spraying and another one is vacuum plasma spraying technique. Smart NiTi alloys which are difficult and expensive to manufacture by conventional methods. Low pressure thermal spraying process technology will reduce lead time, production cost for semi-finished NiTi foils and tubes. The microstructure of the NiTi foils or tubes can be improved by hot rolling .

Thin Film Fabrication

NiTi thin films can be fabricated by using vacuum sputter deposition of multiple layers, micro electromechanical systems (MEMS) and photolithographic steps . Chemical composition of sputtering targets is the primary determinant of phase transition temperatures. In NiTi if titanium possess more composition produces best results, this technique eliminates the difficulties of joining, welding, or bonding of nitinol films. Presently fabrication of medical devices is most successful by planar sputters on three dimensional substrate or multiple layer method. Compositions for Nitinol thin film noted in various studies include binaries in the range of 48.2-51.9 atomic percent Ni, NiTiHf and NiTiCu using either pre-alloyed or elemental targets.

Crystal structures of Shape Memory Alloys

Different metals and alloys will possess different crystal structures and all the metals and their alloys will not show the Shape memory effect. The constitutional changes occurred at the atomic level that contributes the particular properties like Shape memory effect and Superelasticity which are attainable through a solid state phase change having similar that of a molecular re-arrangement but the molecules which remains close packed. SMA can exist in two different phases (Parent phase Austenitic and Daughter phase Martens tic) with three different crystal structures (twinned marten site, de-twinned marten site and austenite) six possible transformations. Parent Austenitic phase is strong having 35000 to 100000 psi yield strength and daughter martensitic phase is fairly weak 10000 to 20000 psi deformation stress able to absorb 8% Recoverable strain. The structure of the high temperature Austenitic phase of Nitinol alloy is B2 type ordered structure and this structure was finally determined by using single X-ray diffraction method and least square analysis. Monoclinic martensite is quite different from martensitic structures in β - phase alloys due to the difference in mechanical properties of high temp Austenite and low temperature martensite. Martensitic transformations are classified into thermo elastic and non-thermo elastic. Non-thermo elastic transformations occur mainly in ferrous alloys thermo elastic martensitic transformations are associated with an interface between austenite phase and martensite phase. According to Otsuka during the rescind transformation there will be shrinkage of martensitic plates will be preferred than nucleation of the high temperature austenitic phase which cause to a crystallographically reversible transformation. The unique properties like Shape memory effect (SME) and super elasticity of SMA results of thermo elastic martensite transformation.

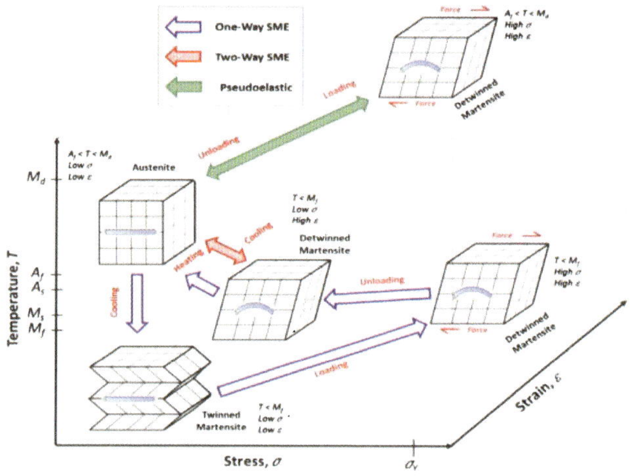

Figure: SMA phases and crystal structures

Properties of SMA

Shape Memory Effect

During martensitic phase transformation the molecular structure is twinned. On a macroscopic scale the size and shape of undeformed martensite phase is same as the cubic austenitic phase. The temperature at which starting and finishing of both parent austenitic phase and daughter martensitic phase has characterized by the following variables M_S, M_f, A_S, A_f. M_S is the martensite start temperature upon cooling and M_f is the martensite finish temperature upon cooling, during heating A_S and Af are the temperatures of the austenite starts and finishes. The loading quantity of SMA increases with the four variables (M_S, M_f, A_S, A_f), shape memory effect (SME) is noticed when SMA temperature is below M_f, when the alloy is in deformed martensite, SMA will be recovered the original shape by heating the specimen above Af.

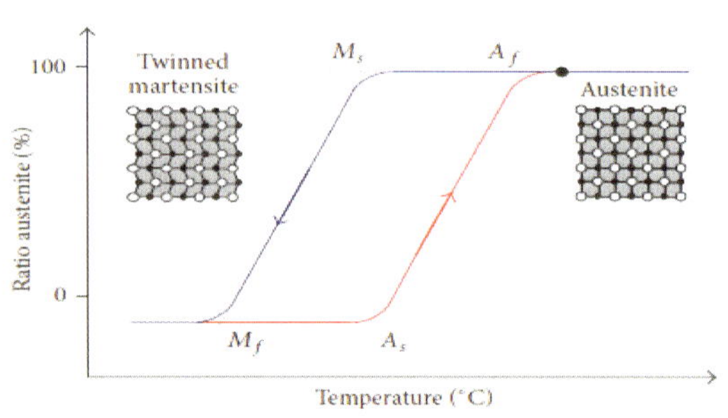

Figure: Temperature Hysteresis in SMA

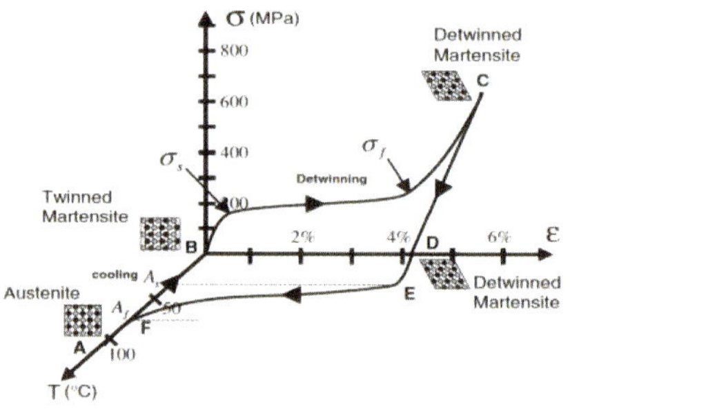

Figure: Stress-strain temperature of NiTi SMA

Normally SME refers to the one-way SME in which external load makes to include de-twinning brings the SMA in to a current distorted structure that can be regain on heating above Af .In this type there is no transformation strains are induced during cooling, Two -way SME is the one in which transformation strains are persuaded during heating or cooling of SMA . Two-way SME is not an essential, but a developed characteristic.

Pseudo Elasticity

Super elasticity or pseudo elasticity of SMA involves stress induced strain recovery upon unloading at a temperature above Af. In general super elastic thermo mechanical loading path starts at zero stress state where de-twinned martensite is stable examples of this characteristic are isothermal (constant temperature) and isobaric (constant pressure) loading paths indicated in. For achieving the required constant stress by austenite is not shown in the constant pressure path. The point which we remember is the constant temperature conditions carry out only by quasi-static loading which we can treat as small strain increments. During phase transformation, there is a dissipation of latent heat which was generated during quasi-static process.

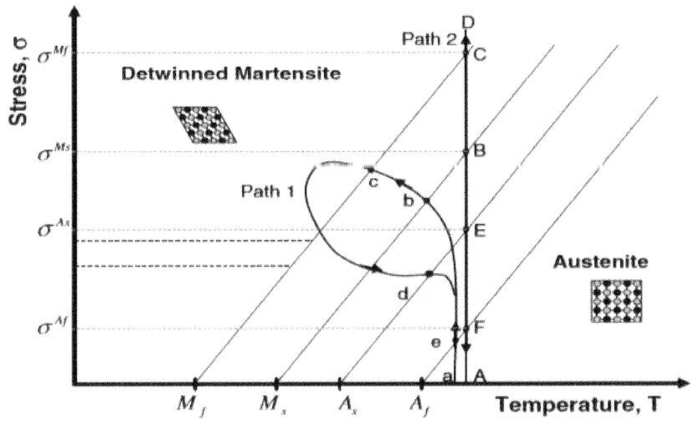

Figure: phase diagram and two possible pseudoelastic loading paths

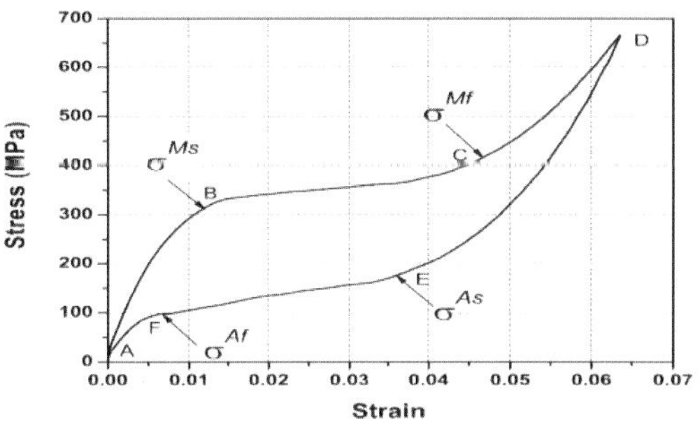

Figure 6 SMA pseudoelastic loading cycle

During loading at a temperature above Mf the transformation occurred at critical stress levels from austenite to martensite that stress is called as Transformation stress (a-b). This phase transformation usually occurs during (b-c) path. During thermo elastic critical stress level large in elastic strains are going to be developed. If the load increased further de twinned martensite region (c-d) does not produce any more phase transformation, during multi-axial loading there may be re-orientation of martensite twins will occur. At point (d) there will be Reverse transformation (RT) from martensite to austenite which will leads to recovery of in elastic strains. At point (e) there will be a complete transformation from martensite to austenite and the final element of the loading path (e-a) is identified by regaining of thermo elastic strains which leads to zero macroscopic strains upon completion of the path. This transformation process concludes in a hysteresis which returns the energy dissipated in the cycle.

Damping Properties

It is very important property for all the materials which are going to use in any specific application. SMA having high damping capacity compared to the remaining materials. Damping property mainly deals with the dissipation of mechanical energy in to heat. The numerous interfaces, which exist during martensitic transformation between austenite and martensite, different martensite variants, the boundaries obtained during twinning of martensite instead of thermo elastic transformation. There will be occurrence of many irreversible events like production defects, dislocation movements etc. hysteresis observed during pseudo elastic characteristic is one of the energy dissipation. The ratio of dissipated energy to the total energy obtained during transformation cycle for isotropic materials will develop. This ratio depends on excitation frequency, temperature, amplitude. A novel material like SMA mainly depends on operating and transformation temperatures.

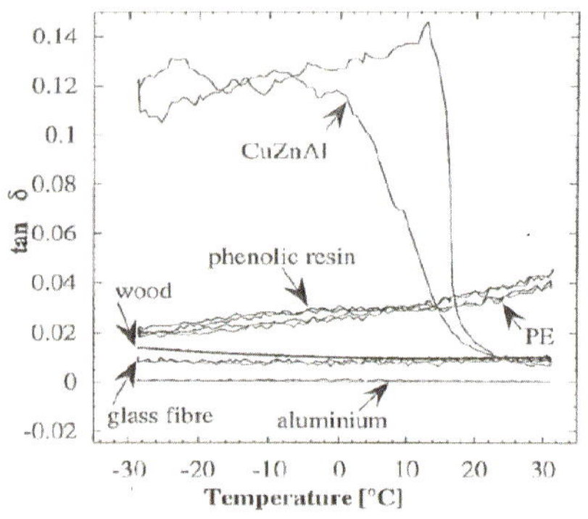

Figure: Damping factor Vs temperature for an frequency of 13Hz and strain amplitude of 10^{-4}

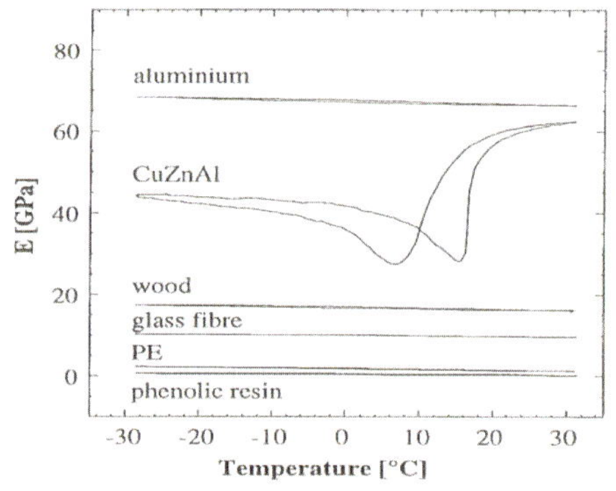

Figure: Elastic modulus vs. temperature for an frequency of 13Hz and strain amplitude of 10^{-4}

The properties of SMA are determined by their composition and the most commercially adopted shape memory alloys are NiTi (Nitinol) and the Cu-based alloys like Cu-Zn-Al and Cu-Al-Ni. Some of the typical properties of these alloys are listed in below table.

	Nitinol (Ni-Ti)	Cu-Zn-Al	Cu-Al-Ni
Melting temperature(^{O} C)	1300	950-1020	1000-1050
Density (gcm^{-3})	6.45	7.64	7.12
Resistivity(μΩcm)	70-100	8.5-9.7	11-13
Thermal conductivity (W cm^{-1}per^{O} C)	18	120	30-43
Young's modulus (Gpa)	83(austenite) 26-48(martensite)	72(beta phase) 70(martensite)	85(beta phase) 80 (martensite)
Yield strength (Mpa)	195-690(austenite) 70-140 (martensite)	350 (beta phase) 80 (martensite)	400 (beta phase) 130 (martensite)
Ultimate tensile strength (Mpa)	895	600	500-800
Shape memory strain (% maximum)	8.5	4	4
Transformation range (^{O} C)	-200-110	<120	<200
Transformation hysteresis (^{O} C)	30-50	15-25	15-20

Behavior and Characterization of SMA

Thermo mechanical characterization of SMA gives qualitative indication of the material behavior and quantitative material properties. Experiments for determining the material property and the behavior has done on the MTS 810 servo hydraulic loading frame Equipment and Differential Scanning Calorimetry (DSC) and has done in four different conditions, resolving of Zero stress transformation, Monotonic loading and unloading, Resolving SME at non zero stress level and finally effects of repeated loading which includes equilibrium of material. Consider a Nitinol wire with a diameter of 1.91 cm used in actuator application, in order to determine zero stress transformation learning of stress free transformation temperatures is necessary otherwise it is very difficult to determine phase of the alloy at any liable stress and temperature. In order to determine these temperatures a DSC was used. The short DSC sample is slit from large quantity material using saw. The test results will obtain after loading the samples in the DSC machine. In order to measure this heat flow/temperature curve tangent to the start and finish of each peak and if we draw tangent to the base line heat will flows. The different intersection of these lines provides measure for start and finish temperatures of each transformation.

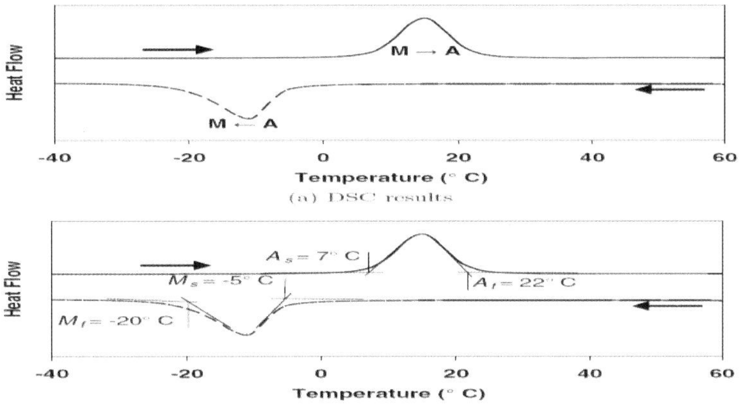

Figure: Determination of stress free transformation temperatures from DSC testing

Consider monotonic loading of SMA if temperature(T)<Martensite finish (Mf) (eg.T<-20° C) gives illumination on elastic properties of twinned martensite also identify if the material indicates essential shape memory behavior by checking the potential to express stress free SME. Before testing, sample is first heated to above Af (eg.22° C) to regain any transformation strain and then cooled to below Mf(- 20° C) to attain a fully martensite state this leads to elimination of detwinned martensite which formed due to unknown deformations. The fragment is consequently tested at a temperature below Mf, Martensite is the only balanced phase. During the course of testing maximum stress is chosen when detwinning has completed. The specimen is then unloaded and the temperature of the specimen is increased homogeneously above Af any recoverable strain occur it is going to be examined. From this test de-twinning start and finish stresses obtained are σs=140Mpa & σs=170Mpa and strain recovered is about 6.2% some non recoverable strain will remain at the end of SME test, elastic modulus is found to be 24 Gpa. If T>Mf (eg.T=30° C) then monotonic loading is applied to the specimen then full super elasticity property is observed with enduring plastic strain. The martensite elastic modulus is found to be higher in this case T>Mf compared with earlier case T< Mf some portion of austenite may remain after the completion of loading. During un loading elastic modulus will reflect austenite and martensite According to ASTM test method F-2516, during loading upper level of stress at 3% strain is 300Mpa and lower level of stress at 2.5% strain is 100Mpa, elastic modulus is found to be 47 Gpa.

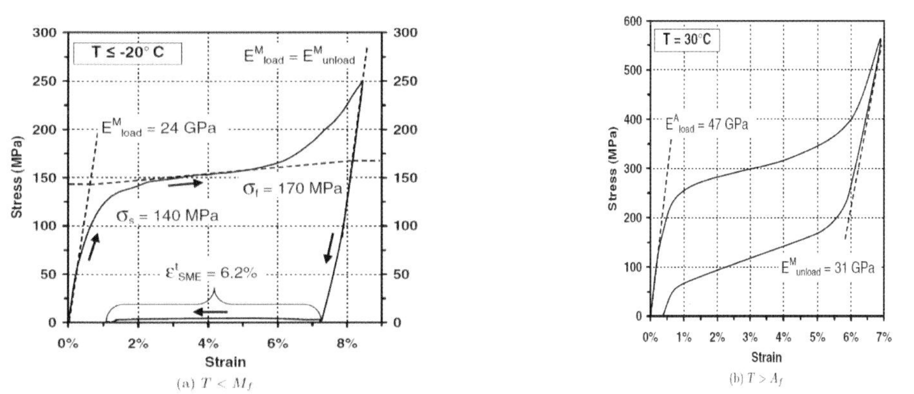

Figure: Results of Monotonic loading/Unloading

During non-zerostress level especially in actuators it is crucial to examine zero SME and material is capable to accomplish work by providing supplanting under load. Constant stress actuation test or isobaric test is going to be adopted for uni-axial loading, for example SMA sample is heated above A_f (22° C) and then stressed to 200Mpa this load held to be constant when the temperature is gradually consistently reduced untill forward transformation in to martensite is completed then temperature is raised gently untill reverse transformation is concluded. Throughout this experiment, the strain is examined and registered.

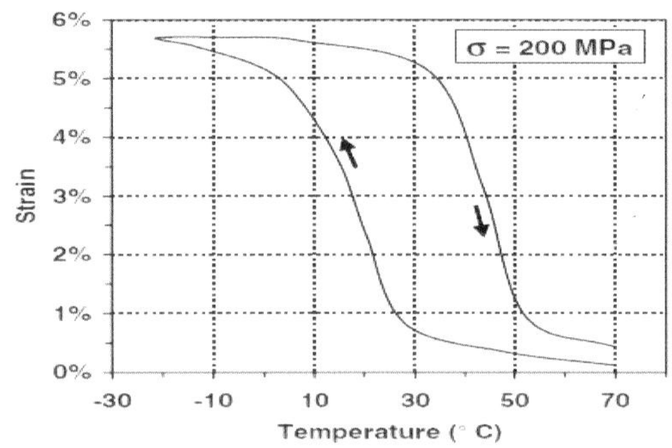

(a)

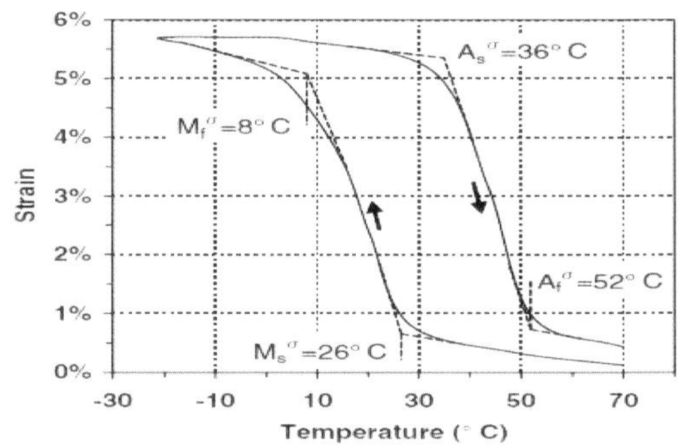

(b)

Figure: strain temperature loading	b) Transformation temperatures underconstant stress thermal cycling Curve underconstant stress thermal cycling

During cyclic loading, when the SMA (eg. actuator) deals with multiple transformation cycles a straight forward approach exists of implementing abundant thermal transformation cycles under steady load can be applied. A constant stress of 200Mpa or below is performed for repeated and reversal loading. Overall of eighty thermal cycles are adopted. When the material is subjected to cyclic loading, its response has emerged and stabilized. Stresses ranges from 2.5Mpa to 200Mpa in 50Mpa accession, the low stress test is important for determining Zero-stress transformation temperaturesand Zero stress transformation strain.

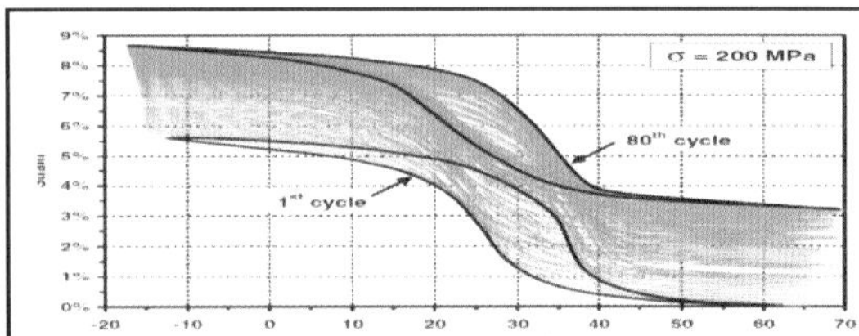

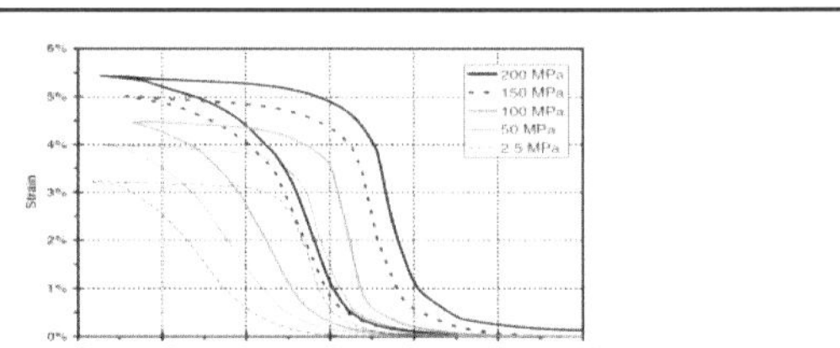

Figure: Results of isobaric cycling at 200Mpa	Results of constant pressure cycling at various stress levels

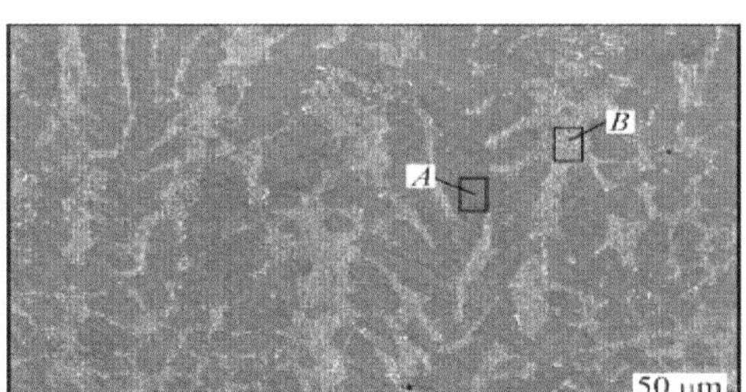

(a) Microstructure of NiTi SMA using SEM

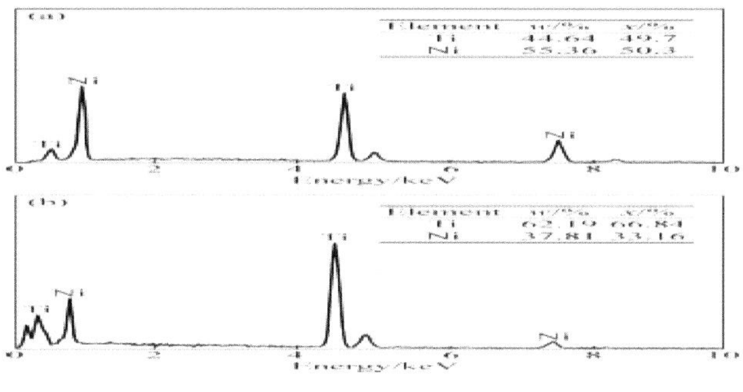

(b) Microstructure of NiTi SMA using EDX Spectra Area A in figure: (a) b) Area B in figure: (a)

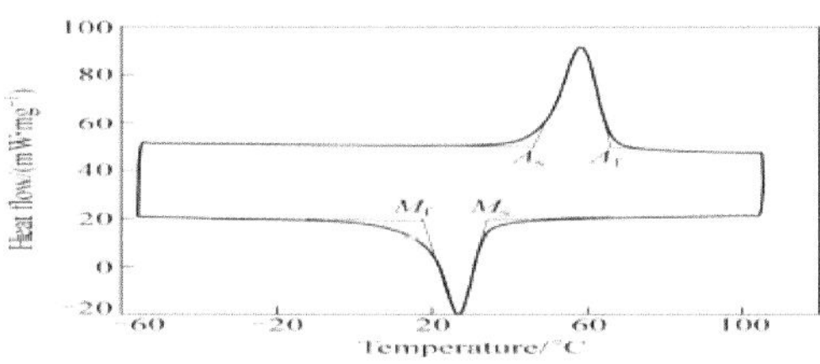

c) DSC of NiTi SMA

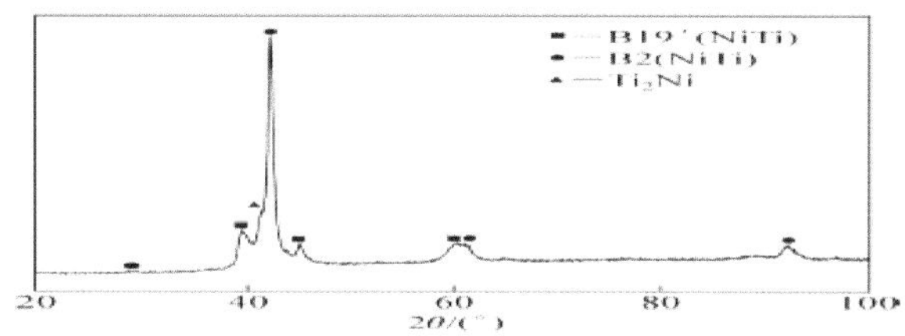

d) Microstructure of NiTi SMA using XRD

Figure (b) shows the EDX spectra of the cast NiTi SMA which exhibits that the matrix phase of the cast NiTi SMA is NiTi and the Ti2Ni phase circulates among the NiTi matrix phases due to this Ti2Ni phase cause the matrix to be rich in Ti. Figure(c)gives DSC curve, the transformation temperatures of the cast NiTi SMA are lowered due to the Ti2Ni phase exists in the NiTi SMA during melting and solidification which increases Ni concentation of the matrix and there by decreasing of transformation temperatures. Figure (d) gives XRD analysis of cast NiTi SMA consists of B19' martensite phase B2 Austenite phase & Ti2Ni phase simultaneously at room temperature which is in concord with the DSC test. Martensite & Austenite co-exist because the Mf of the cast NiTi SMA is smaller than the room temperature, which causes insufficient transformation of cast NiTi SMA from Austenite to Martensite at room temperature

Applications of SMA

In Automobile industries SMA plays a major role. SMA used as actuators, sensors, mini actuators. Automatic valves, we can also use SMA as Solenoids in order to reduce noise and electric generators to generate electricity from exhaust heat. It can also used as air dams to optimize aerodynamics at various speeds. NiTi SMA can be used as a variable area form nozzle which makes jet engines quite and efficient, vibration dampers in jet engines, high shock absorber applications such as ball bearings, & landing gears. SMA having an advantage of reducing weight and increasing efficiency. In Robotics SMA used as Micro actuators or artificial muscles the difficulties come across in biomedical fields, small strain output, low efficiency, large hysteresis, slow response times.SMA find varieties of applications such as buildings,& bridges, Intelligent Reinforced Concrete (IRC) is one of the SMA application uses wires embedded with the concrete also it senses cracks and reduce large scale sized cracks. Due to biocompatibility property, SMA used in orthodontic devices, endodontic files, broken bones can have remedy with SMA. In order to prevent scalding SMA can make to restrict water flow by reactors at different temperatures.SMA can also adopted in petro chemicals, semi conductors, pharmaceuticals, oil industries, piping, Engines, crafts, Gas boilers. Applications which deal with damped vibrations, free recovery, wire and ribbon applications, heated fluid flowing through tubes, SMA are ideal. SMA plays a very important role in commercial applications which includes eye glass frames. SMA can also be used as a material for helicopter blades which depends mainly on vibration in Micro processing control tabs for the trailing ends of the blades.

Future Trends of SMA

Developments of new SMA have significantly improves the quality of SMA attributes and performances. A new concept called as multiple memory material technology (MMMT) which was developed by researchers at university of waterloo, Canada has transformed SMA in to multiple shapes at various temperatures. Current research focuses on developing iron based shape memory alloys having thermo elastic martensitic transformation are necessary to get shape memory effect (SME).many alloys like Fe-Ni-C, Fe-Mn-Si, Fe-Mn-Si-Cr-Ni, undergoes non thermo- elastic transformation and still exhibits good SME. These alloys are having different characteristics compared with conventional SMA in that they depends on stress-induced martensite for SME also it offers large transformation hysteresis, less (4%)recoverable strain commercial capability of these alloys has still to be determined but the effort has opened up for new classes of alloys which includes β-Ti alloys and iron base alloys. The importance of modern computer design and analysis tools such as CAD & FEA in to innovative and a reliable SMA application which makes interest for accurate 3D constitute models. Development of new materials which includes composites and hybrid shape memory materials, fabrication technologies and treatment process which makes stable, durable and also develops robust computation models of SMA behavior. Linear actuator was developed by Velázquez et al. has used mini actuator made of two bias springs has a diameter of 0.15cm and a length of 4.5cm, a weight of 150 milli grams and is capable of developing a draw force of 320 mN. Donnellan suggested mini linear actuator which was developed by Starsys Research and Applied physics Laboratory (ALP) for getting larger stroke instead of using SMA wire if we use SMA compression spring of 7.6mm length and 6mm diameter we can able to get larger stoke of 1.7 cm with an alert time of 1 sec. Elwaleedet. al has also developed SMA beam actuator to amplify the actuator strain using elastic beams. SMA wire of diameter of 0.7 mm 150mm gives a displacement up to 2cm which is greater than the unconstrained wire movement. Pittaccio has developed the mini linear SMA actuator which is adopted for cyclic and apathetic mobilization of in elastic ankle generates a maximum movement of 8cm. Park et al has developed Mini-rotary- actuators using Nitinol SMA wires having a diameter of 100μm which changes completely from treading rotational motion to the extended rotational motion. Linear actuators in the stroke to volume ratio (SR)-force to volume ratio (FR) plane can trace a straight line through the location of Donnellan's actuator and parallel to the one representing the ideal SMA mini actuator characteristics. Figure 19 can differentiate in to three areas, the line gives the present implicit technological restriction for developing mini actuators. Below this restriction nearly all the actuators has conferred with in this area with little SR and FR. Suppose in this area if there is a big SR and little FR is very limitedly explored as yet i.e. only Haga's actuator fall into this region. The third area is presently un-explored, For examining this area for developing SMA mini actuators with mechanical accomplishment that suits high FR and low SR. A more intricate and challenging imminent goal is to bring close to the SMA mini actuator mechanical accomplishment by changing implicit technological restriction near to the ideal one.

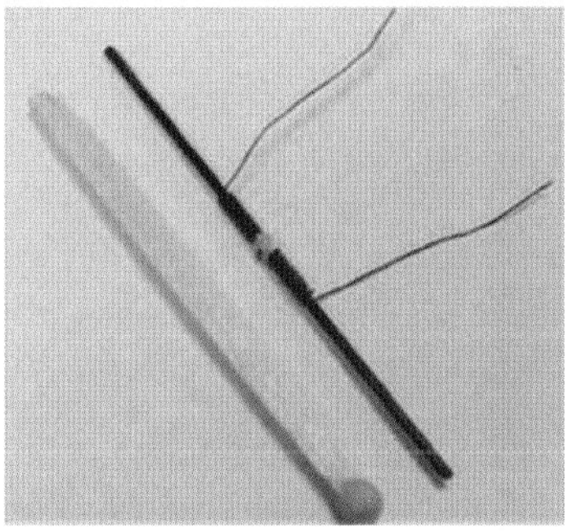

Figure: Mini actuator Proto type

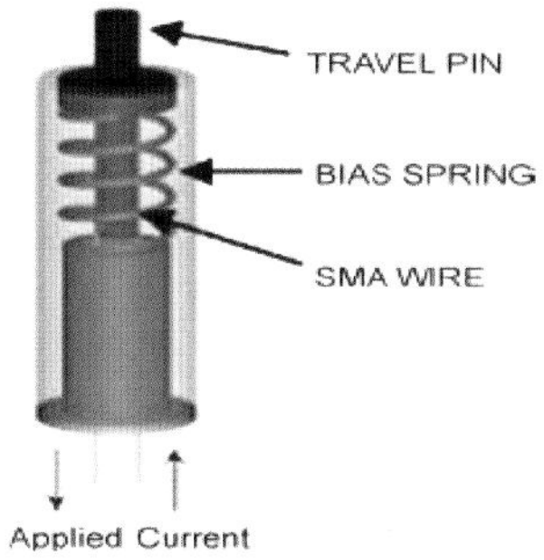

Figure: Starsys Research and Applied Physics

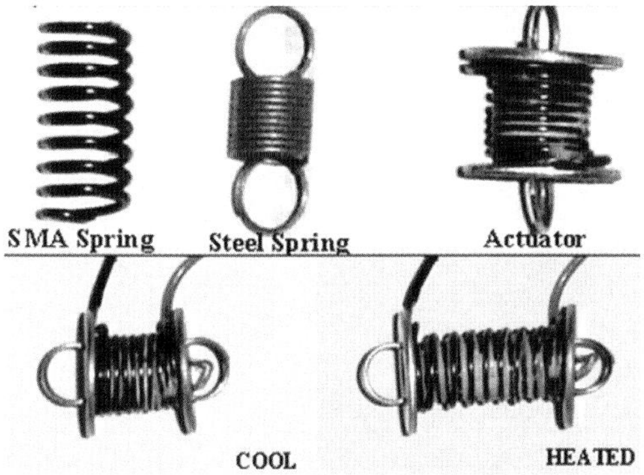

Figure: Improvement of the mini-linear-actuator

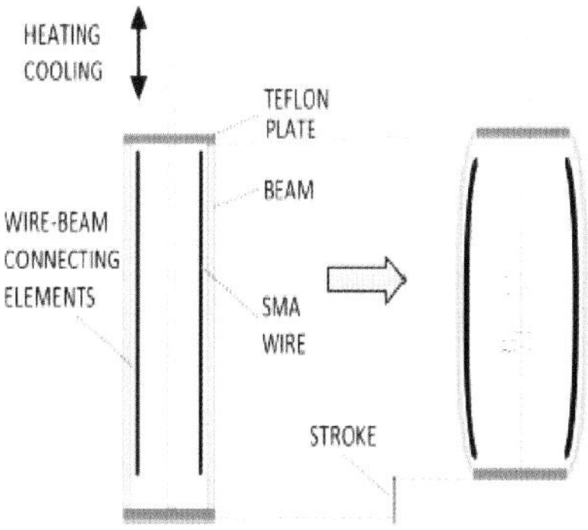

Figure: actuator Representation.

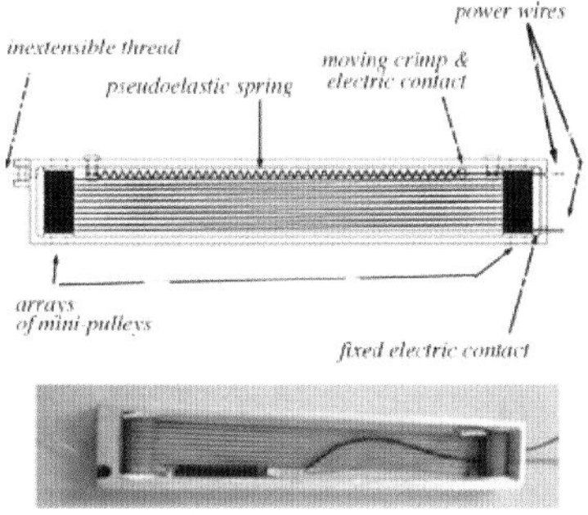

Figure: representation and prototype of. Mini linear actuator

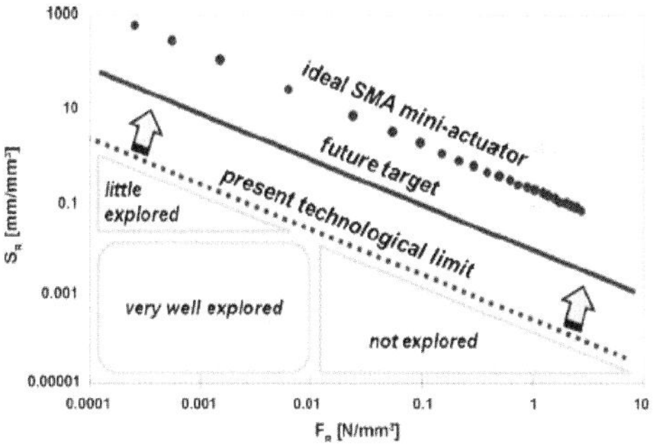

Figure: Future aspect on SMA mini actuators

References

- Ham, Peter "Zinc amalgam" in e-EROS Encyclopedia of Reagents for Organic Synthesis (2001). doi:10.1002/047084289X.rz003
- Metal-alloys-2340254: thebalance.com, Retrieved 10 April 2018
- "Die Sendung mit der Maus, Sachgeschichte vom Spiegel" (in German). Archived from the original on 17 April 2009. Retrieved 2009-04-24.
- Alloy-definition-examples-and-uses-606371: thoughtco.com, Retrieved 29 May 2018
- Amorphous-Metal-Alloys: appropedia.org, Retrieved 30 March 2018

- Mutter, Joachim (13 January 2011). "Is dental amalgam safe for humans? The opinion of the scientific committee of the European Commission". Journal of Occupational Medicine and Toxicology. 6: 2. doi:10.1186/1745-6673-6-2. PMC 3025977. PMID 21232090..
- Difference-between-substitutional-and-interstitial-alloys: pediaa.com, Retrieved 19 July 2018
- Emmanuil I. Troyansky and Meghan Baker "Aluminum Amalgam" in e-EROS Encyclopedia of Reagents for Organic Synthesis 2016, doi:10.1002/047084289X.ra076.pub2
- Faq-what-is-an-oxide-dispersion-strengthened-ods-alloy, technical-knowledge: twi-global.com, Retrieved 19 June 2018

5

Metallurgical Processes

In order to completely understand metallurgy, it is necessary to understand the processes related to it, such as Bayer process, Bessemer process, Betts electrolytic process, patio process, Pidgeon process, Hoopes process, etc. which have been thoroughly discussed in this chapter.

Bayer Process

Alumina is the main components of bauxite, so bauxite is refined in order to produce Alumina. The Bayer process is the principal way for producing alumina by refining Bauxite. Bauxite other than with 30-60percentage of aluminium oxide contains mixture of silica, iron oxides and titanium dioxide. The Bayer Process is the most economic means of obtaining alumina from bauxite. Other processes for obtaining alumina from metal ores are also in use in some refineries, particularly in China and Russia, although these make up a relatively small percentage of global production. Prior to the Bayer process bauxite is crushed and ground in mills to fine particles (max. size0.06”/1.5mm). A hot solution of the recycled sodium hydroxide (caustic soda, NaOH) is then added to the ground ore forming a bauxite slurry, which is stored in holding tanks and then pumped to the further processing stage.

The Bayer process can beconsidered in three stages:

BAYER PROCESS FLOW SHEET

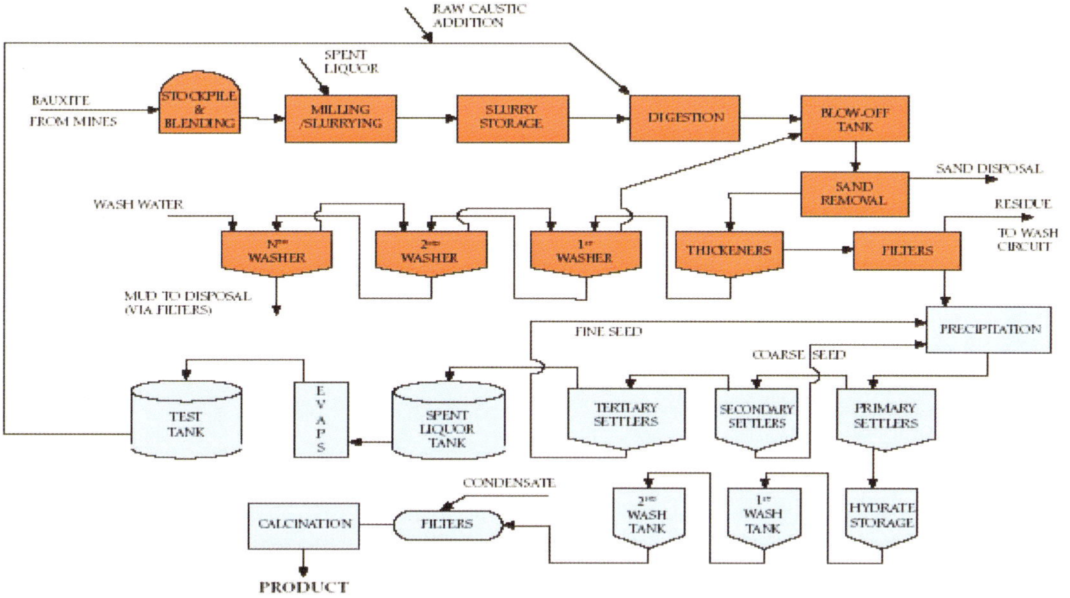

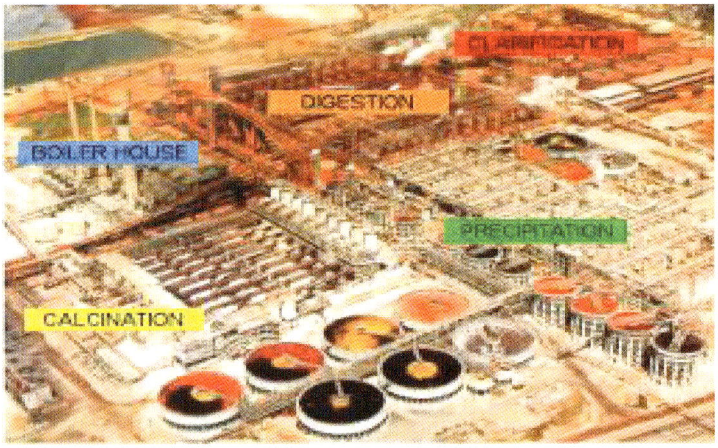

Figure: Aerial view of a Bayer processing plant

Extraction

The aluminium-bearing minerals in bauxite - Gibbsite, Böhmite and Diaspore -are selectively extracted from the insoluble components (mostly oxides) by dissolving them in a solution of sodium hydroxide (caustic soda):

The first step in the process is the mixing of ground bauxite into a solution of sodium hydroxide. By applying steam and pressure in tanks containing the mixture, the bauxite slowly dissolves. The alumina released reacts with the sodium hydroxide to form sodium aluminate.

Gibbsite:

$$Al(OH)_3 + Na^+ + OH^- \rightarrow Al(OH)_4^- + Na^+$$

Böhmite and Diaspore:

$$AlO(OH) + Na^+ + OH^- + H_2O \rightarrow Al(OH)_4^- + Na^+$$

Depending on the quality of the ore it may be washed to beneficiate it prior to processing. The ore is crushed and milled to reduce the particle size and make the minerals more amenable to extraction. It is then combined with the process liquor and sent in a slurry form to a heated pressure digester.

Conditions within the digester (concentration, temperature and pressure) are set according to the properties of the bauxite ore. Ores with a high Gibbsite content can be processed at 140° C with about 110gl^{-1} of Na_2O. Processing of Böhmite on the other hand requires harsher conditions, between 200 and 240°C and 140-170 gl^{-1} of Na_2O. The pressure is not important for the process, as such but is defined by the steam pressure during the actual process conditions. At 240°C the pressure is approximately 35 atmospheres (atm).

Although higher temperatures are often theoretically advantageous there are several disadvantages including corrosion problems and the possibility of oxides other than alumina dissolving into the caustic liquor, leading to impurities problems in the final product, and scaling of pipes, pumps and tanks.

After digestion about 30% of the bauxite mass remains in suspension as a thin red mud slurry of silicates, and oxides of iron and titanium. The mud-laden liquor leaving the digestion vessel is flash-cooled to atmospheric boiling point by flowing through a series of flash vessels which operate at successively lower pressures.

The flash steam generated is used to preheat incoming caustic liquor in tubular heat exchangers located parallel to the flash tank line. Condensate from the heat exchangers is used for boiler feed water and washing waste mud.

After the extraction stage the insoluble bauxite residue must be separated from the Aluminium-containing liquor by a process known as settling. The mixture of solid impurities is called red mud, and presents a disposal problem. The liquor is purified as much as possible through filters before being transferred to the precipitators. Slaked lime is added to dilute caustic liquor in the washing process to remove carbonate (Na2CO3) which forms by reaction with compounds in bauxite and also from the atmosphere and which reduces the effectiveness of liquor to dissolve alumina. Lime regenerates caustic soda, allowing the insoluble calcium carbonate to be removed with the waste mud.

The insoluble red mud from the first settling stage is thickened and washed to recover the caustic soda, which is then recycled back into the main process.

The mud is washed with fresh water in counter-current washing trains to recover the soda and alumina content in the mud before being pumped to large disposal dams.

Precipitation

Crystalline aluminium trihydroxide (Gibbsite), conveniently named "hydrate", is then precipitated from the digestion liquor:

$$Al(OH)_4^- + Na^+ \rightarrow Al(OH)_3 + Na^+ + OH^-$$

This is basically the reverse of the extraction process, except that the product's nature is carefully controlled by plant conditions, including seeding or selective nucleation, precipitation temperature and cooling rate. The cooled pregnant liquor flows to rows of precipitation tanks which are seeded with crystalline trihydrate alumina, usually of an intermediate or fine particle size to promote crystal growth. Each precipitation tank is agitated, with a holding time of about three hours. During the 25-30 hours pass through precipitation, alumina of various crystal sizes is produced. The entry temperature and the temperature gradient across the row, seed rate and caustic concentration are control variables used to achieve the required particle size distribution in the product.

The finished mix of crystal sizes is settled from the liquor stream and separated into three size ranges in three stages "gravity" classification tanks. The primary classifiers collect the coarse fraction which becomes the product "hydrate". The intermediate and fine crystals from the secondary and tertiary classifiers are washed and returned to the precipitation tanks as coarse and fine seed. The process engineer decides on the ratio of coarse and fine seed to be added, based on screen analyses.

The slurry of coarse "hydrate" ($Al_2O_3.3H_2O$) from the primary thickeners is pumped to hydrate storage tanks and is filtered and washed on horizontal-table vacuum filters to remove and recycle process liquor, prior to calcination in fluid bed or rotary calciners. Heating the slurry before filtration, with waste steam, gives a cleaner, drier filter cake and leads to less soda loss. The heated cake entering the calciner also helps fuel efficiency.

Calcination

"Hydrate", is calcined to form alumina for the aluminium smelting process. In the calcination process water is driven off to form alumina, this takes place at 1050° C:

$$2Al(OH)_3 \rightarrow Al_2O_3 + 3H_2O$$

The calcination process must be carefully controlled since it dictates the properties of the final product. Sandy alumina particles are 90%+ 45 µm (microns) in size, this is the product sought after by most smelters.

A large amount of the alumina so produced is then subsequently smelted in the Hall-Héroult process, named after inventors Charles Hall and Paul-Louis Héroult, in order to produce aluminium.

Smelting

The purified Al2O3 is dissolved in molten cryolite, Na3AlF6, which has a melting point of 10120C and is an effective conductor of electric current. In the following schematic diagram of the electrolysis cell graphite rods are employed as anodes and are consumed in the electrolysis process. The cell electrolytic reaction is:

$$2Al_2O_3 + 3C \rightarrow 4Al(l) + 3CO_2(g)$$

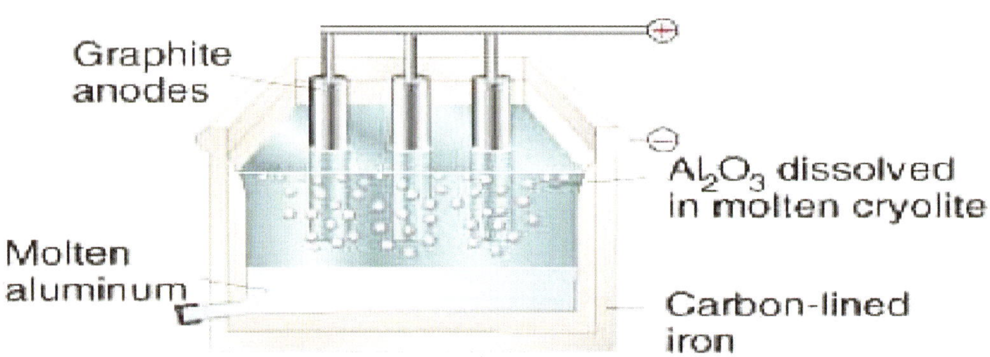

The production of one ton of aluminum requires about 65-70 GJ (18-20 MWh) and about half a ton of carbon.

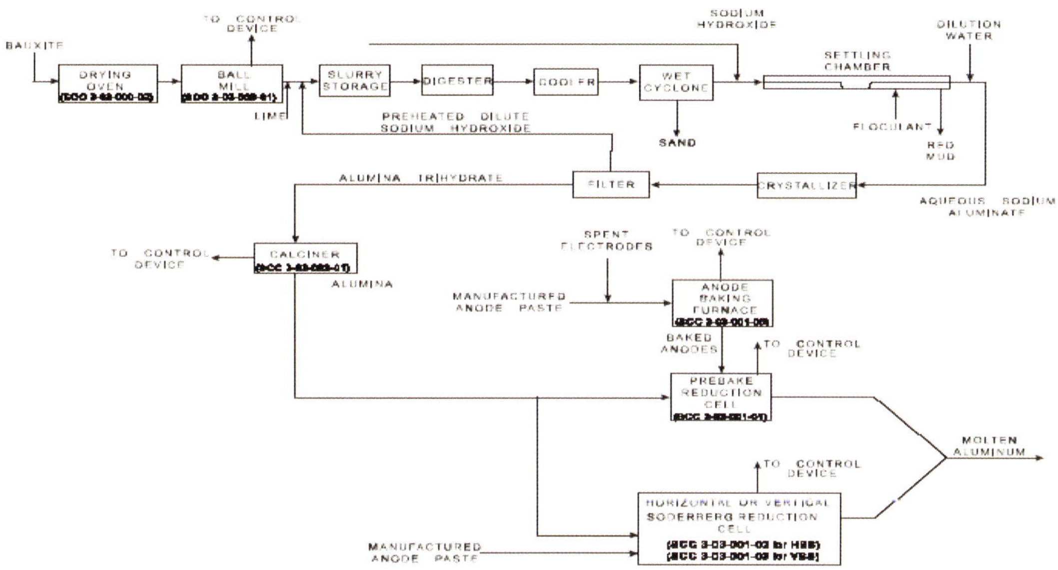

Figure: Schematic diagram of aluminium production

Bessemer Process

The Bessemer process was the first inexpensive industrial process for the mass-production of steel from molten pig iron. The process is named after its inventor, Henry Bessemer, who took out a patent on the process in 1855. The key principle is removal of impurities from the iron by oxidation with air being blown through the molten iron. The oxidation also raises the temperature of the iron mass and keeps it molten.

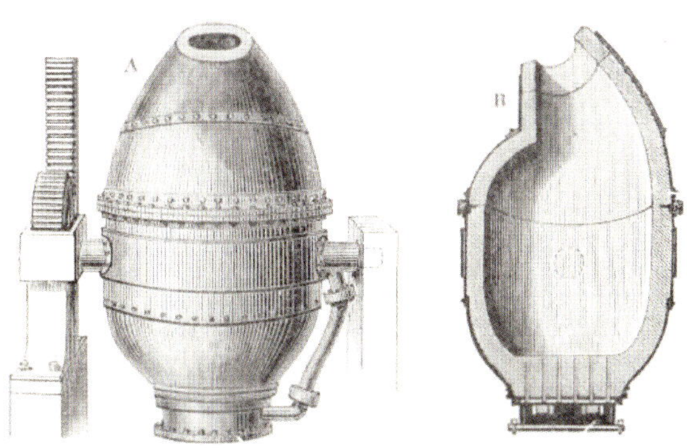

The process is carried on in a large ovoid steel container lined with clay or dolomite called the Bessemer converter. The capacity of a converter was from 8 to 30 tons of molten iron with a usual charge being around 15 tons. At the top of the converter is an opening, usually tilted to the side relative to the body of the vessel, through which the iron is introduced and the finished product removed. The bottom is perforated with a number of channels called tuyères through which air is forced into the converter. The converter is pivoted on trunnions so that it can be rotated to receive the charge, turned upright during conversion, and then rotated again for pouring out the molten steel at the end.

Oxidation

The oxidation process removes impurities such as silicon, manganese, and carbon as oxides. These oxides either escape as gas or form a solid slag. The refractory lining of the converter also plays a role in the conversion—the clay lining is used in the acid Bessemer, in which there is low phosphorus in the raw material. Dolomite is used when the phosphorus content is high in the basic Bessemer (limestone or magnesite linings are also sometimes used instead of dolomite). In order to give the steel the desired properties, other substances could be added to the molten steel when conversion was complete, such as spiegeleisen (an ironcarbon-manganese alloy).

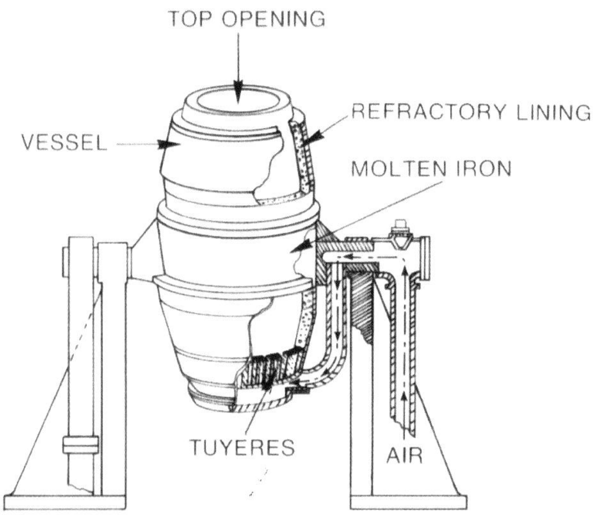

Managing the Process

When the required steel had been formed, it was poured out into ladles and then transferred into moulds and the lighter slag is left behind. The conversion process called the "blow" was completed in around twenty minutes. During this period the progress of the oxidation of the impurities was judged by the appearance of the flame issuing from the mouth of the converter.

After the blow, the liquid metal was recarburized to the desired point and other alloying materials are added, depending on the desired product.

Before the Bessemer process, Britain had no practical method of reducing the carbon content of pig iron. Steel was manufactured by the reverse process of adding carbon to carbon-free wrought iron, usually imported from Sweden. The manufacturing process, called cementation process consisted of heating bars of wrought iron together with charcoal for periods of up to a week in a long stone box. This produced blister steel. Up to 3 tons of expensive coke was burnt for each ton of steel produced. Such steel when rolled into bars was sold at £50 to £60 a long ton. The most difficult and work-intensive part of the process, however, was the production of wrought iron done in finery forges in Sweden.

This process was refined in the 1700s with the introduction of Benjamin Huntsman's crucible steel-making technique, which added an additional three hours firing time and required additional large quantities of coke. In making crucible steel the blister steel bars were broken into pieces and melted in small crucibles each containing 20 kg or so. This produced higher quality crucible steel but increased the cost. The Bessemer process reduced to about half an hour the time needed to make steel of this quality while requiring only the coke needed to melt the pig iron initially. The earliest Bessemer converters produced steel for £7 a long ton, although it initially sold for around £40 a ton.

Sir Henry Bessemer described the origin of his invention in his autobiography. at the time of the outbreak of the Crimean War many English industrialists and inventors became interested in military technology and Bessemer himself developed a method for grooving artillery projectiles so that they could spin without the use of rifling in the bore of the gun. He patented this method in 1854 and began developing it in conjunction with the government of France. After a successful day of testing of his method at the Polygon in France he had a conversation with Claude Etienne Minié who stated that a key barrier to the use of the larger, heavier spinning projectiles would be the strength of the gun and in particular "he [Minié] did not consider it safe in practice to fire a 30-lb. shot from a 12-pounder castiron gun. The real question, he said, was; Could any guns be made to stand such heavy projectiles?". This is what started Bessemer thinking about steel.

At the time steel was difficult and expensive to make and was consequently used in only small items like cutlery and tools. Starting in January 1855 he began working on a way to produce steel in the massive quantities required for artillery and by October he filed his first patent related to the Bessemer process.

Bessemer first started working with an ordinary reverbatory furnace but during a test a couple of pig ingots got off to the side of ladle and were sitting above it in the hot air of the furnace. When Bessemer went to push them into the ladle he found that they were steel shells: the hot air alone had converted the outer parts of the ingots to steel. This crucial discovery led him to completely redesign his furnace so that it would force high-pressure air through the molten iron using special air pumps. Intuitively this would seem to be folly because it would cool the iron, but due to exothermic oxidation both the silicon and carbon react with the excess oxygen leaving the surrounding molten iron even hotter, facilitating the conversion to steel.

Bessemer licenced the patent for his process to five ironmasters, for a total of £27,000, but the licences failed to produce the quality of steel he had promised and he later bought them back for £32,500. He realized the problem was due to impurities in the iron and concluded that the solution lay in knowing when to turn off the flow of air in his process; so that the impurities had been burnt off, but just the right amount of carbon remained. However, despite spending tens of thousands of pounds on experiments, he could not find the answer.

The simple, but elegant, solution was first discovered by English metallurgist Robert Forester Mushet, who had carried out thousands of scientifically valid experiments in the Forest of Dean. His method was to first burn off, as far as possible, all the impurities and carbon, then reintroduce carbon and manganese by adding an exact amount of spiegeleisen. This had the effect of improving the quality of the finished product, increasing its malleability - its ability to withstand rolling and forging at high temperatures and making it more suitable for a vast array of uses.

The Dowlais Iron Company was the first licensee of the Bessemer process, constructing the world's most powerful rolling mill in 1857, and producing its first Bessemer steel in 1865.

The Bessemer process revolutionized steel manufacture by decreasing its cost, from £40 per long ton to £6-7 per long ton during its introduction, along with greatly increasing the scale and speed of production of this vital raw material. The process also decreased the labour requirements for steelmaking.

Prior to its introduction, steel was far too expensive to make bridges or the framework for buildings and thus wrought iron had been used throughout the Industrial Revolution. After the introduction of the Bessemer process, steel and wrought iron became similarly priced, and most manufacturers turned to steel. The availability of cheap steel allowed large bridges to be built and enabled the construction of railways, skyscrapers, and large ships.

Other important steel products were steel cable, steel rod and sheet steel which enabled large, high-pressure boilers and high-tensile strength steel for machinery which enabled much more powerful engines, gears and axles than were possible previously. With large amounts of steel it became possible to build much more powerful guns and carriages, tanks, armored fighting vehicles and naval ships. Industrial steel also made possible the building of giant turbines and generators thus making the harnessing of water and steam power possible. The introduction of the large scale steel production process perfected by the Englishman Henry Bessemer paved the way to mass industrialization as observed in the 19th-20th centuries.

Obsolescence

Commercial steel production using this method stopped in Workington in 1974. It was replaced by processes such as the basic oxygen (Linz-Donawitz) process, which offered better control of final chemistry.

The Bessemer process was so fast (10–20 minutes for a heat) that it allowed little time for chemical analysis or adjustment of the alloying elements in the steel. Bessemer converters did not remove phosphorus efficiently from the molten steel; as low-phosphorus ores became more expensive, conversion costs increased. The process permitted only limited amount of scrap steel to be charged, further increasing costs, especially when scrap was inexpensive. Use of electric arc furnace technology competed favorably with the Bessemer process resulting in its obsolescence.

A slag ladel outside Workington Steel Works in 1986. Up until the 1960's slag from blast furnaces was taken to the shore and dumped into the sea.

Betts Electrolytic Process

In Betts process lead bullion is purified by transferring most of the lead from a soluble anode to a cathode through a lead containing electrolyte while leaving behind impurities in an adherent anode slime. The electro- chemistry of this process involves the properties of anode slimes and of entrained electrolyte. Thus, the processes of anodic corrosion of lead and transport of lead ions through the entrained electrolyte are essential to the understanding of the Betts process.

Extractive Metallurgy of Lead

Lead is an ancient metal. It was used by the Romans for components of their water distribution systems because of its malleable and ductile properties. The ancients made lead of acceptable purity probably by smelting lead ores under conditions that prevented arsenic, antimony, and other hardening elements from reducing to the metallic phase. This could be accomplished in most cases by producing high lead slags such as those still produced in fire as saying. In those days it was necessary to avoid excessive copper in the ore, but arsenic, antimony, bismuth, nickel, iron, etc. were reliably held in the slag by maintaining the high oxidizing conditions of lead silicate based slags. Lead recoveries were tow not better than 85% from the highest grade handpicked galena ore. When high lead recoveries were found to be obtainable by coke-based blast furnace reduction, lead so produced was too hard, usuaily because of its copper, antimony, and arsenic content. The function of lead refining became both a softening process and a method of recovering silver and gold.

Nowadays the extraction process can be conveniently portrayed in the two step flow sheet shown in figure below. The first step involves bullion production from the sulphide concentrate and the second step the refining of bullion to the final product.

The conventional route for bullion production requires sintering of the concentrate to produce a lead oxide containing product, which is then reduced in a blast furnace with metallurgical coke to produce lead bullion. The KIVCET and QSL processes represent two relatively recent commercial developments that replace both the sintering-blast furnace combination with a single furnace that treats concentrates, and reduce both the costs of lead smelting and the environmental impact.

The KIVCET process replaces the sintering/blast furnace operations with a flash smelting step. In this process, lead sulphide is oxidized to lead bullion and sulphur dioxide in a stream of oxygen. In a second step, the bullion and slag flow under a weir to an electrically heated settling hearth where coke breeze or coal is added to reduce the residual lead oxide in the slag and produce a final bullion (for refining) as well as a low-lead stag.

The QSL process consists of a long, horizontal, tubular, brick-lined converter in which lead concentrates are pelletized and injected near one end into a bath containing lead bullion, lead oxide-containing slag, and lead sulphide matte. Oxygen is blown into the bath in the feed injection (and lead-bullion discharge) zone where it ultimately oxidizes sulphide sulphur to sulphur dioxide gas. Slag is tapped from the far end of the reactor after passing through a zone where reducing coal-air mixtures are injected through tuyeres to lower its lead oxide content.

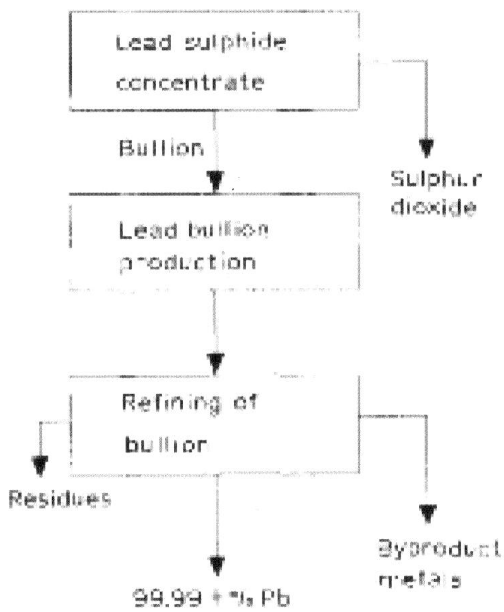

Figure: Flow sheet for lead extraction from sulphidic concentrates.

The refining of bullion is carried out by either a pyrometallurgical route or a combined pyrometallurgical/electrometallurgical route. Comparisons between these two routes show that the pyrometallurgical route is usually used when ores with low bismuth content are treated. The generalized pyrometal- lurgical flowsheet is shown in figure below. This process consists of a series of steps which capitalize on a complex series of phase relationships to extract all the impurities contained in the lead bullion down to very low levels.

The combined pyrometallurgical/electrometallurgical route is shown in figure below. Here, copper drossing is performed to remove the bulk of the copper as a combination of matte and arsenide-antimonide for further treatment, thus allowing for the removal of some arsenic and antimony. Arsenic and antimony are sometimes reduced further as sodium arsenate- antimonate dross by oxidizing in the presence of caustic soda, because their levels in bullion must be controlled to produce suitable anodes for successful electrorefin- ing practice.

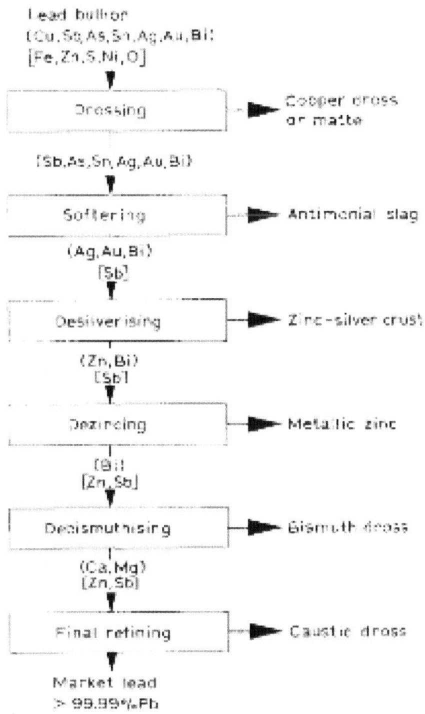

Figure: "A generalized flow sheet for the pyrometallurgical refining of lea d. Parentheses ( ) indicate a major impurity component. Brackets [ ] indicate a minor impurity component.

Electrorefining is carried out in either a fluosilicic, fluoboric or sutphamic acid electrolyte and produces a commercial lead cathode product and an anode with an adhering slime. The purity of the produced lead is usually higher than 99.99%. The slimes, representing only 2 to 4% of the anode weight, are treated by a variety of processes to recover silver, copper, antimony, gold, bismuth, and sometimes tin and indium.

Processes that entirely avoid smelting (and its attendant gas and dust treatment systems and associated environmental risks), utilizing hydrometallurgical/ electrometallurgical flowsheets, have also been proposed to replace the current technology. These include (a) the US Bureau of Mines ferric chloride leach process which recovers lead via the molten salt electrolysis of $PbCI_2$, (b) the Minemet Recherche ferric chloride leach process which recovers lead from chloride leach solutions using aqueous electrolysis and (c) the US Bureau of Mines process for leaching lead concentrates in waste fluosilicic acid with an oxidant (hydrogen peroxide or lead peroxide) followed by aqueous electrolysis from the fluosilicic acid leach solution. All these processes have been piloted, but none has been commercialized.

Plant Practice in Lead Electrorefining

The BEP process is used in Canada, China, East Germany, Italy, Japan, Peru, Rumania, Russia, USA and West Germany. The average annual production of lead by BEP is approximately 1 000 000 tons. Since the production of refined lead in the non-socialist countries (for which good figures are available) is close to 4 700 000 tons per year, it can be assumed that up to about 20% of world lead production is refined by the BEP. Two variations of the BEP, the sulphamic acid and the ftuoboric acid processes are operated in Italy and in West Germany respectively.

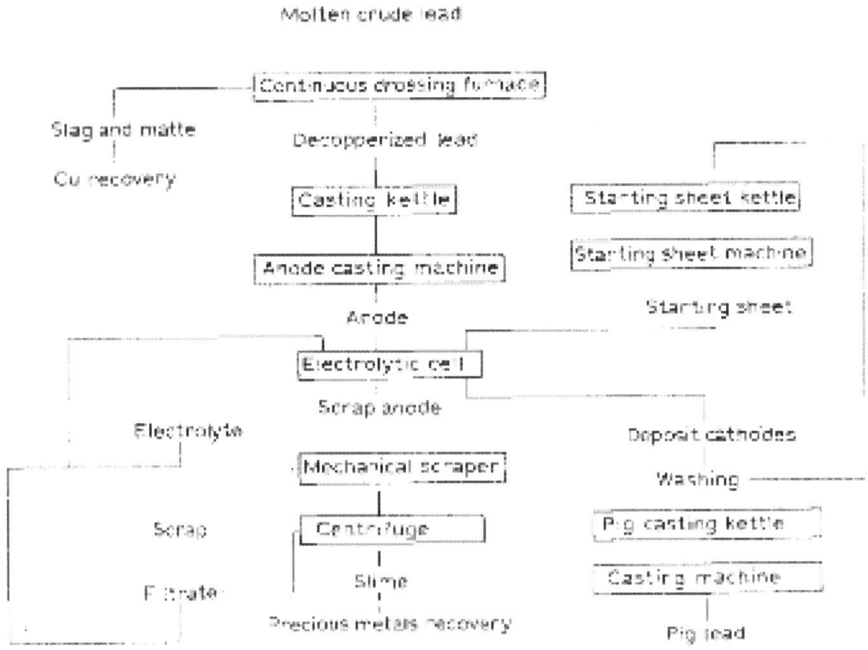

Figure: A generalized flowsheet for the pyrometallurgical/electrometalturgical refining of lead.

The Betts Electrorefining Process normally utilizes a fluosilicic acid (H2SiF6) electrolyte containing lead ttuositicate to electrorefine impure lead anodes into pure lead cathodes. The fluoborate and the sulphamate processes are identical to the fluosilicate process except for the substitution of the electrolyte. Fluoboric acid has not been extensively used due to its relatively high cost. HBF4 is a stable acid with a good electrolytic conductivity and a high solubility of the lead salt. It is also used in lead plating baths, lead fluoborate- fluoboric acid rechargeable batteries and in the recovery of lead from spent batteries.

The sulphamic acid process has two main limitations when compared to the fluosilicic acid process: firstly, sulphamic acid decomposes rapidly at current densities larger than 100 A m^{-2} resulting in high reagent replacement cost; secondly, the free acid is a crystalline solid of limited solubility, and limited ionization in aqueous solution, leading to a low con- ductivity of sulphamic acid solutions (at most, half of the conductivity of equivalent fluosilicic acid solutions). This results in larger power costs in refining. The significant advantage of the sulphamic acid process is that it is the most efficient process for the removal of tin from impure lead (Sn remains in the slimes). Plants that use the fluosilicic process remove Sn prior to electrolysis by using the Harris process. When anodes containing significant Sn concentrations are refined by the Betts process, tin dissolves with lead at the anode and co-deposits at the cathode. The lead-tin alloy may be sold, or some pos t treatment of the cathodes is necessary to remove the tin.

The wide use of H_2SiF_6 in lead electrorefining is due to its low cost. Fluosilicic acid is produced as a by-product of the treatment of phosphate rock in fertilizer manufacture. Fluorides and silica containing in phosphate rock form fluosilicate and are separated from the fertilizer product. Fluosilicic acid is also produced as a by-product during the dissol- ution of apatite with sutphuric acid.

During the electrorefining operation, fluosilicic acid is con- sumed by entrapment in the anode slimes and by volatilization from the surface of the electrolyte. The acid develops significant vapour pressures through volatile decomposition products according to the reaction,

$$H_2SiF_6 \rightarrow 2HF(g) + SiF_4(g)$$

SiF_4 and HF are both corrosive and toxic and are removed from the tankhouse atmosphere by adequate ventilation.

Higher current densities have been achieved through the use of galvanodynamic techniques such as current modulation and periodic current reversal (PCR). The current modulation technique consists of decreasing the current density (for example, from 220 to 160Am- 2) in small steps. Each constant current density step is determined on thebasis of an anode over potential value which increases as the slimes layer thickens and decreases when the current density is reduced. The upper limit for the anode over potential is usually determined by bismuth dissolution from the anode. Below figure shows the current density program that can be applied to the electrorefining circuit without reaching the critical over potential value for bismuth dissolution. The higher average current density possible with current modulation reflects in a shorter electrorefining cycle and higher refinery production.

Periodic current reversal (PCR) in lead electro- refining is widely employed in China. PCR involves frequent short reversals of the electrolysis current direction. This reduces the concentration polarization in the slimes layer and levels the cathodic deposit by selectivity dissolving projections. High current efficiencies, good cathode quality, low electrolyte losses, low energy consumption and a decrease in the number of short circuits has been reported through the use of PCR. A 16% increase in free acid was found in the slimes layer due to PCR.

Anodic Process

The Betts process, as practiced by all refineries, depends on the formation of an adherent, porous anode slimes layer during electrolysis. The slimes layer consists of undissolved impurities which are removed mechanically from anode scrap after the anodes are withdrawn.* If the slimes do not adhere to the anode 95'.

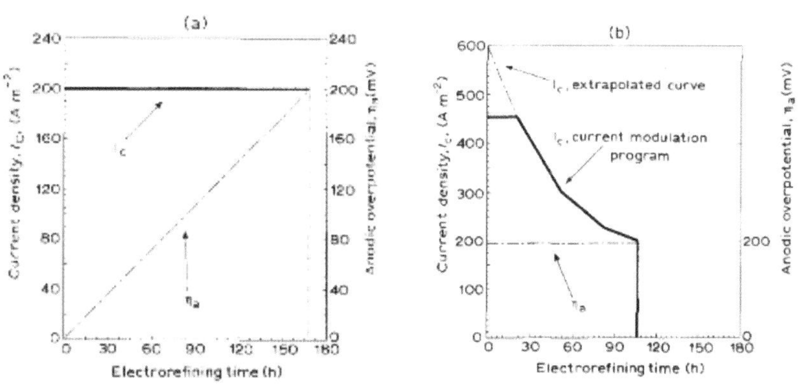

Figure: Changes in the anodic over potential value, η_a during lead electrorefining: (a) conventional galvanostatic process (b) galvanodynamic process in which the cell current is continuously decreased during the refining cycle to fix the value of η_a.

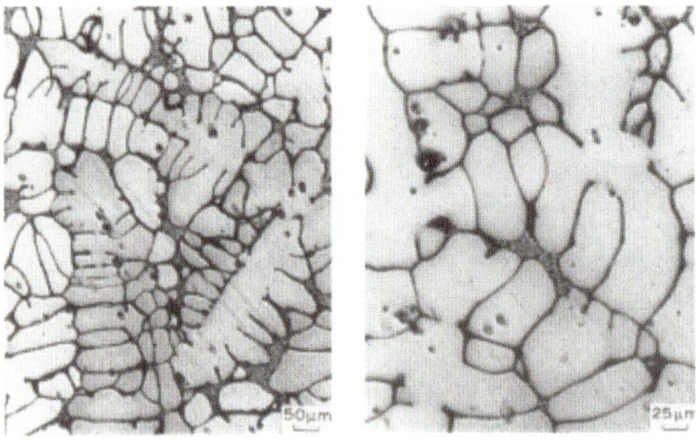

Lead anode microstructure. From an anode currently being used by Cominco Ltd. Chemical composition as described in Table1. Top view, air cooled side.

Refining Of Lead by the Betfs Process

During electrolysis they will settle through the solution and to some extent be mechanically entrained in the cathode deposit. There is no indication in the literature of a lead refining process in which (as in copper refining) slimes t:atl is encouraged. Further, the recovery of slimes from the bottom of the cell (as in copper refining) is perceived to be more costly. To form an adherent slimes layer, certain elements (mainly As, Sb, and Bi) must be controlled within a narrow composition range and/or ratio in the anode, and the anode casting process must be designed to control the rate of solidification to optimize the microstructure of the cast bullion.

Physical Metallurgy of the Lead Anodes

The lead anode microstructure that is desired for optimum slimes structure during electrorefining is known as the honeycomb structure, because it consists of uniform size grains of lead surrounded by impurities on the grain boundaries. As the lead grains dissolve they leave behind a skeleton of slimes which resembles a honeycomb. Any non-uniformity of the matrix will lead to non-adherence of the slimes layer, and any precipitates present in this matrix material will contribute to slimes detachment because of their extra weight.

There are four elements present in the lead anodes that seem to exert a strong influence on the anodic process: As, Sb, Bi and Ag. The interaction of these impurities with each other and with lead can be deduced from the available binary and ternary phase diagrams [89-93], which show that both intermetallic compounds and eutectic structures may be present. Three different slime forming systems have been identified and classified from studies on synthetic anodes:

(i) Impurity phases of the solid solution type (SST): the Pb-Bi system.

(ii) Impurity phases of the precipitation type (PT): the Pb-Sb system.

(iii) Impurity phases of the eutectic type (ET): the Pb-As and the Pb-Ag systems.

The strength of slimes adhesion to the anode was quantified by Tanaka through observations of slimes fall and slimes morphology. Tanaka summarized the results of these studies with the following relationships:

(1) The greatest slimes adherence is obtained when impurity phases of the SST type are present in the anode in concentrations greater than 0.23% by wt. A SST concentration lower than this critical value produced a slime that easily slides off the anode.

(2) The addition of a third element to the eutectic systems increases significantly the slimes adherence.

(3) Water quenching of the anodes increase slimes adhesion particularly in the PT and ET system cases. However, the very small slimes particles formed during quenching may also promote slimes detachment and mechanical entrainment in the cathodes.

The increase of slimes adhesion by water quenching relates directly to the lead anode solidification rate, which influences the growth and distribution of impurity-containing phases. Especially important are the eutectic forming systems where it is known that the solidification parameters (growth velocity, temperature gradient in the liquid and growth mechanism) and system parameters (volume fraction and impurities content) can lead to anomalous eutectic structures. Such structures have been reported to occur in the following systems:

Ag-Pb and Ag-Bi: broken lamellar structure type

Pb-Bi: complex structure type

Pb-Sb: complex regular structure and irregular structure type.

It must be emphasized that even though the physical metallurgy of the lead anodes is of great importance to the Betts process, there remains a lack of knowledge in this area.

Industrial Practice

The range of anode impurities used in lead refining varies from plant to plant. In every refinery the anode composition is kept within narrow limits to obtain an adherent slimes layer. The need for production of anodes with homogeneous properties has led to control methods for cooling rates and casting techniques. In addition to the use of a casting wheel, straight horizontal and vertical anode casting systems are employed,

The straight type horizontal lead anode casting system currently used in Japan. Although the vertical and horizontal casting processes were originally developed to save space (over that occupied by a casting wheel), they were carefully designed to achieve uniform cooling rates during the casting of the anodes. Such new techniques for anode casting as well as for optimizing heat treatment have been patented by Japanese companies , but no information was found that would indicate the nature of changes in microstructure resulting from these newer techniques.

Slimes Electrochemical Behavior

The slimes layer formed in the BEP undergoes structural and chemical changes during its growth. As the slimes layer thickens, the different phases and com- pounds present react with the entrained electrolyte creating secondary products. The Eh-pH diagram for the $PbSiF_6$-H_2SiF_6 system indicates that both lead fluoride and silica can precipitate at higher pH values. This has been related to the effect of electrolysis parameters on secondary processes (such as these precipitations) that take place within the slimes layer during electrolysis. The transport processes within the slimes layer in the context of these secondary reactions has also been the subject of several studies.

A simple physical model to study the role of the slimes layer during anodic dissolution rate in electro- refining systems was developed by Reznichenko. In this model, the slimes layer was represented by a silver gauze diaphragm (representing the noble impurity) electrically connected to the anode (a highly pure base metal) and placed at a (variable) distance from the anode corresponding to the slimes - electrolyte interface. The distance between the anode and the gauze was varied to simulate the effect of an increasing ohmic drop between the anode and the slimes layer. During electrolysis, concentration gradients normally present in a slimes layer were replaced by a simple concentration difference between the solutions on the two sides of the gauze, and this was related to a modified Nernst equation:

$$\text{In}\left[\text{Me}_2^{n_2^+}\right]=\left(E_1^{\circ}-E_2^{\circ}+\Delta\phi+\frac{RT}{n_1F}\text{In}\left[\text{Me}_2^{n_1^+}\right]\right)\frac{n_1F}{RT}$$

where: $\left[\text{Me}_2\right], n_2$ = concentration and valence of noble metal; $\left[\text{Me}_1\right], n_1$ = concentration and valence of base metal; $\Delta\phi$ = ohmic drop between the base metal and the noble metal gauze; E_1° = rest potential of the metal to be refined; E_2° = est potential of the noble metal; F = Faraday's constant; R = gas con- stant; and T = temperature.

Equation $\text{In}\left[\text{Me}_2^{n_2^+}\right]=\left(E_1^{\circ}-E_2^{\circ}+\Delta\phi+\frac{RT}{n_1F}\text{In}\left[\text{Me}_2^{n_1^+}\right]\right)\frac{n_1F}{RT}$ shows that the larger the ohmic drop the larger the equilibrium concentration of ions of the noble metal and the larger its dissolution rate. Although this model oversimplifies the different phenomena taking place within a slimes layer, it pro- rides an insight as to the effect of ohmic drop buildup in the slimes layer and the dissolution rate of noble impurities normally left in the slimes. Reznichenko. applied their model to the study of the Cu-Ag system in which they found a logarithmic relationship between the dissolution rate of noble impurities and the ohmic drop potential.

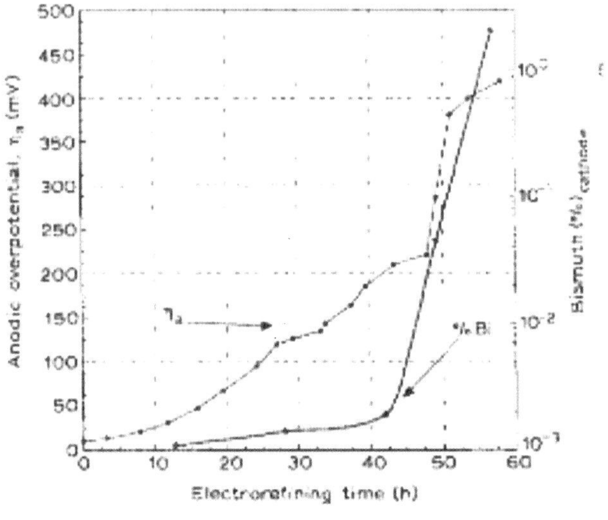

Figure: Bismuth content of the cathodic deposit as a function of the anodic polarization and the electrolysis time. Ic = t70-350 A m -2. Anode composition 8.4-11.2% Bi, remainder Pb.

Early research on the behavior of BEP slimes has focused on relating the anodic overpotential to the cathode purity. Among the impurities most closely followed in the cathodic deposit is bismuth. Even though the BEP has a large selectivity for the removal of bismuth, it has been found that towards the end of the electrorefining cycle such selectivity can be lost. It seems that the buildup of concentration gradients and the precipitation of secondary products within the slimes layer can initiate steep increases or discontinuities in the anodic over- potential at a certain slimes thickness. This would produce a large increase in the rate of dissolution of such impurities, and subsequently in the deposition of impurities in the cathodic deposit. Figure above shows how bismuth contamination in the cathode increases with anodic overpotentials above a critical value (about 200 mV). This has been recognized as a general behavior for bismuth in lead refining by the BEP. On the other hand, the use of periodic current reversal (PCR) has been found to increase the minimum anodic overpotential at which noble impurities start to dissolve.

The slimes layer weight is only 1 to 4% of the original anode weight; yet, they occupy the whole of the original anode volume. This indicates that the porosity of the slimes layer exceeds 92% and may be as high as 98% (taking into account a density for slimes phases of about half that of the bullion). Concentration gradients in the electrolyte confined within this highly porous layer and the slimes electrochemistry are closely linked.

Wenzel measured the change in composition of the electrolyte contained in the slimes layer as a function of time and of slimes thickness. Figure shows the sampling method used by Wenzel, utilizing sampling wells at different distances from the anode surface. Small amounts of electrolyte were withdrawn through these wells at carefully selected times (to avoid perturbing appreciably the system).

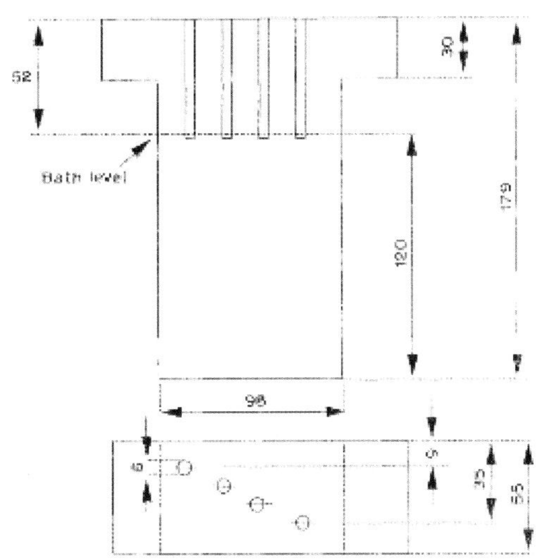

Figure: Dimensions of the anodes used in Wenzel's experiments showing the position and size of the electrolyte sampling wells. All measurements in ram.

According to their results, using a range of anode compositions (Bi from 0 to 1.74% and Sb from 0.45 to 3.01%), the more the amount of secondary solidified material (SSM) present in the anode, the steeper the pH and the Pb^{+2} concentration gradient throughout the slimes layer. Also they found that the thicker the slimes layer the steeper the concentration gradients. Figure shows these changes in concentration for two different anode compositions. Wenzel et al. proposed that due to the pH increase several reactions may occur: SiF_6^{-2} decomposition to SiF_4 and F (pH > 3); precipitation of Sb2O3 (pH > 4.9) and precipitation of Pb_2O (pH > 7).

Table below show the results of chemical and diffraction analysis, which seems to indicate the presence of these secondary compounds within the slimes layer*. From an analysis of the concentration of noble impurities in the entrapped electrolyte, Wenzel were able to conclude that when the SSM exceeded 5%, as determined from the Pb-Bi-Sb ternary diagram, permissible impurities in the anode were too high to obtain an acceptable impurity level in the refined lead.

Wenzel's X-ray diffraction analysis on the slimes pro- ducts agree with the findings of Isawa et al., who studied the slimes composition of synthetic and industrial lead anodes and found the presence of the following compounds:

Metallic Bi: in the Pb-Bi, Pb-Bi-As, Pb-Bi-Sb, and Pb-Bi-As-Ag systems;

Metallic As: in the Pb-As, Pb-Bi-As, and Pb-Bi- As-Ag systems;

Metallic Ag: in the Pb-Ag and Pb-Bi-As-Ag systems.

Chemical composition and X-ray diffraction analysis of the lead anode slimes

Sample no.	Anode composition (% wt) (rest Pb)	Slimes chemical analysis (% wt)					X-ray diffraction analysis
		Pb	Sb	Bi	F	Si	
1	0.46% Sb, 0.24% Bi	56.98	7.62	3.95	11.97	6.66	PbO, Sb_2O_3, PbF_2, Bi_2O_3
2	0.46% Sb, 0.24% Bi	58.02	9.40	4.28	14.09	3.33	PbO, Sb_2O_3, PbF_2, Pb_4SiO_4, Bi_2O_3
3	0.92% Sb, 0.24% Bi	56.57	6.81	5.55	11.27	6.65	PbO, Sb_2O_3, PbF_2, Bi_2O_3
4	0.92% Sb, 0.24% Bi	53.77	9.37	9.20	12.22	4.19	PbO, Sb_2O_3, PbF_2, Bi_2O_3

Also the presence of the Bi-Sb solid solution, of the ε and ε' phases of the Ag-Sb system, and of water soluble As and Sb (identified as As2O3 and of Sb2O3) were found in some of the above mentioned systems and in the slimes obtained from industrial anodes.

One compound that has been difficult to characterize is Ag3Sb , whose presence has been detected in both synthetic and industrial lead anodes. The morphology of this compound has been found to be a function of the anode cooling rate and of the electrolysis conditions.

The electrolyte used in lead electrorefining contains impurities whose concentration ranges from a few ppm and several gl^{-1}, and have their origin in the anode from which they are dissolved (i.e. C_u +2, S_n +2) or from the acid manufacturing process (i.e. phosphorous species). Their steady state concentrations in the electrolyte depend on the electrolysis parameters and slimes thickness. They can affect both the anodic and cathodic processes through changes in fundamental electrochemical parameters such as the exchange current density (i_0), the transfer coefficient (α), and the electrical double layer capacity (e.d.1.). Measurement of these parameters usually involve transient electrochemical techniques, such as polarization scans, cur- rent interruptions, and AC impedance studies. Miyashita have extensively studied the influence of minor impurities and addition agents in the electrolyte used in lead refining. These studies focused on the determination of fundamental electro- chemical parameters (i_0, e.d.1., and α) as a function of Sn^{-2},Fe^{+3},Zn^{+2} and glue, using a single current step transient method. He found that under an oxygen atmosphere the presence of the above ionic species increased the anodic overpotential value. Thus, they may inhibit the dissolution of some noble species including lead.

When noble impurities present in the lead anodes transfer to the electrolyte they can be redeposited by cementing on less noble elements. While such redeposition is obscured during anodic dissolution, it can be inferred from corrosion measurements in the absence of a current. For example, the weight loss rate for a lead sample in the presence of dissolved bismuth has been observed in the HBF_4-$Pb(BF_4)$ 2 system. Figure below shows how bismuth concentrations as low as 2 mM enhance the corrosion rate of lead.

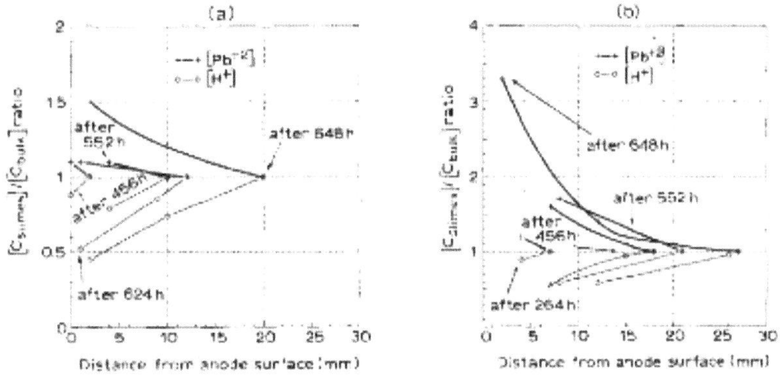

Figure: Changes with electrolysis time of the relative concentrations of Pb^{+2} and H + within the slimes layer with respect to their bulk values.

Bulk electrolyte composition: [$PbSiF_6$] = 0.4 M, [H2SiF6] = 1 M, T = 35° C, I = 200 a m -2, stationary electrodes. Vertical axis: Pb $^{+2}$ and H^+ concentration ratio between the electrolyte sampled within the slimes layer and the bulk electrolyte. (a) Anode with 0.46% Sb and 0.24% Bi (~ 1.73% SSM), (b) anode with 0.92% Sb and 0.24% Bi (3.80% SSM). Note: no extrapolation of the inner electrolyte concentrations at zero slimes thickness done here.

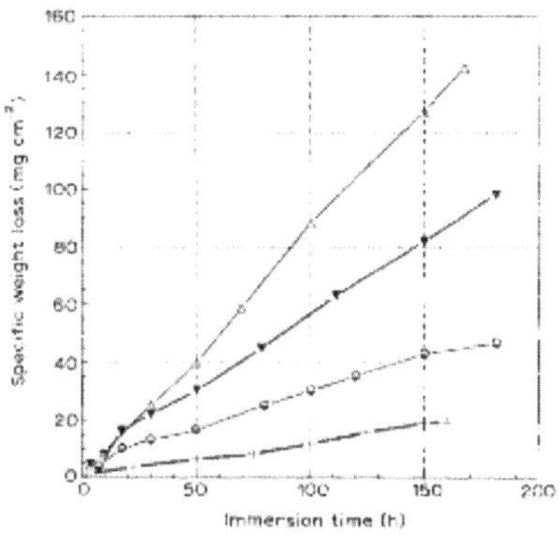

Figure: Lead specific weight loss (corrosion) as a function of the immersion time in a 2 M Pb $(BF_4)_2$ - l M HBF 4 solution in the presence of(+) 0, (Θ) 2, (∇) 5, and (Δ) 10Mm1^{-1}ofBi $^{+3}$.

During the dissolution of the lead anodes in the BEP, bismuth and other noble impurities accumulate in the proximity of the slimes-bulk electrolyte interface as well as in the bulk electrolyte. Since these noble impurities will deposit on the cathode at their limiting mass transfer rate, the permitted electrolyte concentration is related to cathode purity specifications (25 mg 1-i of bismuth will typically lead to 10 ppm bismuth in the cathode). They can be removed by one of two methods:

(a) Continuous cementation through a column filled with lead particles.

(b) Electrodeposition in a separate electrolysis circuit. The purified electrolyte is then sent to the main electrolyte stream. This method is being used in Japan.

The majority of plants that electrorefine lead do not purify their electrolyte, but rather, operate under conditions where the dissolution of noble impurities from the anode is limited to tolerable levels.

Cathodic Process

Additives Control and Electrochemistry

In the absence of additives, lead deposits with a very small overpotential, and tends to form rough, porous deposits or dendrites that result in short circuits. To obtain deposits that are flat, smooth, and free from projections, special reagents are added to the electrolyte. These 'addition agents' increase the cathodic overpotential (actually called 'inhibition'), and change the kinetic parameters ($i_{o,\alpha}$ e.d.l.) under comparable electrolysis conditions.

Polarization measurements have been used for con- trolling and monitoring the concentration of additives in lead electrorefining circuits, as welt as for screening of additives in the electrolytic refining of copper and for testing of impurity levels in the electrowinning of zinc. A polarization technique for determining lignin sulphonate in lead plating baths has also been described. Long term studies and years of industrial practice, especially by Cominco researchers have lead to a sufficient under- standing of the levelling mechanisms of lignin su!- phonate and aloes, to permit the use of polarization measurements for controlling the concentration of these species in the industrial electrolyte for lead refining. This additives control is achieved by keeping the cathode polarization voltage (c.p.v.) within preset limits, and adjustments are made by either new additions or by regeneration (in the case of aloes) with a thiosulphate salt. The level of additives present at a given time is directly related to the c.p.v.

Starting Sheet Technology

The cathode starting sheets used in the BEP are usually cast on a rotary drum casting machine. The stiffness of these sheets is increased by adding approximately 15 ppm antimony to the melt and by impressing Continuous purification of electrolyte via purification coaumn wrinkles on the cathodes right after they are produced. Procedures to avoid dross incorporation in the lead cathodes have also been developed. In this topic of the electrorefining process high levels of automation have been achieved.

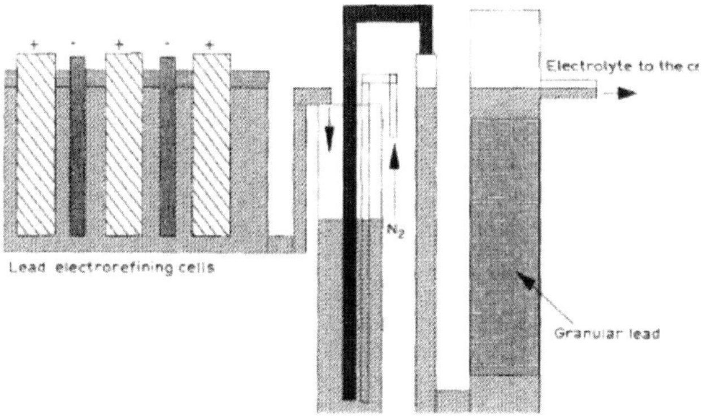

Figure: Continuous purification of electrolyte via purification columm.

Cell Electrolysis Parameter Optimization

The overall electrorefining process has been studied using statistical correlations of some of the variables measurable in an operating plant. To optimize the electrolysis parameters when the BEP is run at high current densities $\left(\gg 200Am^{-2}\right)$ actorial design of experiments at three levels for four variables has been used by Lange. They varied PbSiF6 and HaSiF6 concentrations, temperature, and current density, at constant values of addition agent (glue), anode composition, cell geometry, and electrolyte recirculation. The influence of these parameters on the average anodic and cathodic polarization, on the cell voltage, and on the specific energy consumption was obtained by employing regression analysis correlations. In addition, changes in these parameters were correlated with the cathode quality and appearance as well as with the consumption of glue. The optimum electrolysis parameters found for a current density increase from 200 to 300 Am^{-2}) were a Pb +2 concentration of 60 g 1^{-1} (down from 80 g 1^{-1}), a glue addition maxi- mum of 1500gton of lead (up from $1250g1^{-1}$), a free H_2SiF_6 content of $120gl^{-1}$ (no change) and an electrolyte temperature of 35 to 40° C. Under these electrolysis conditions, Lange et al. found that antimony and bismuth slimes layers could be subjected to anodic polarizations as high as 280mV without serious impairment of the cathode quality.

Optimization of the BEP through half cycle anode slimes scrubbing and cathodes exchange was also investigated by Lange. Not only did the specific energy consumption decrease by the implementation of this procedure, but highly pure cathodes were assured by avoiding anodic overpotentials in excess of the preset limits for impurities dissolution.

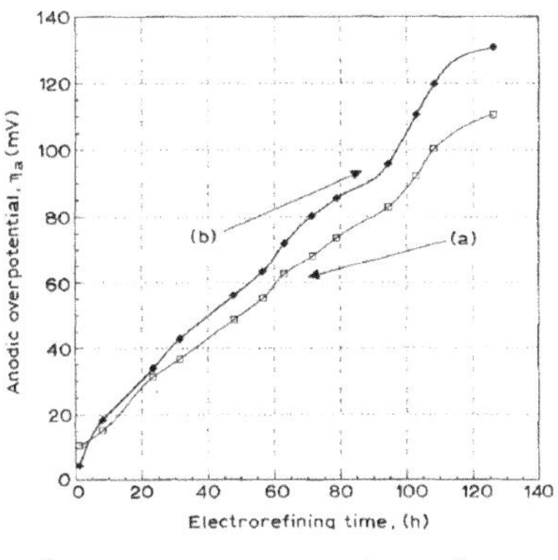

Figure: Influence of the cycle length and current reversat ratio on the anodic overpotential value during PCR: $(a)T_{\text{forward}} / T_{\text{backward}} = 11.5, T_{\text{cycle}} = 10\text{s}$. $(b)T_{\text{forward}} / T_{\text{backward}} = 19, T_{\text{cycle}} = 20\text{s}$.

On the other hand, the exchange of cathodes and the half cycle scrubbing of the anodes incorporates labor increases making implementation difficult. The use of this technique is necessary when anodes containing high impurity levels (> 5%) are to be treated. For example, in Russia, lead anodes containing as much as 15% bismuth are scrubbed every 48 h during the 6 day long anode cycle. In addition to this, cathodes are exchanged daily to avoid short circuits and to obtain a deposit of acceptable purity.

Lange et al. also studied the use of PCR in the BEP. Anodic overpotential dependance on the PCR parameters. Thiet found that the use of PCR decreased the anodic overpotential values. The reduction in the anodic overpotential values did not reflect in lower PCR energy consumption probably because it was offset by unproductive energy consumption during the reverse current phase of the cycle. Thiet also showed that the quality of the PCR cathode deposits was very high.

Bipolar Refining of Lead

Among other alternatives to optimize the lead electro- refining process, the use of a bipolar configuration looks most promising. In this configuration only the terminal electrodes are connected to the source of current. Between these electrodes a large number of bipolar electrodes can be incorporated. One side of these electrodes will corrode anodically, while pure lead will deposit on the other side. The impurities will be left behind as a slimes layer on the anodic side. The advantages of operating the BEP in a bipolar mode include PCR, c.p.v., optimum lead and free acid con- tents, and the use of jumbo electrodes (~ 4 m 2, possible because there are no bus-bar connections)*. The process has been operated in a pilot plant where it proved to be superior to the parallel process. The rationale for the current use of the parallel system is that a high economic investment is required for the substitution and implementation of the bipolar process. If new plants to electrorefine lead are to be constructed in the future, very likely they will be assembled on the bipolar configuration.

Patio Process

Patio process is also called Mexican process. It is the method of isolating silver from its ore that was used from the 16th to early in the 20th century; the process was apparently commonly used by Indians in America before the arrival of the Europeans.

The silver ore was crushed and ground by mule power in arrastras, shallow circular pits paved with stone. Large blocks of stone attached by beams to a central rotating post were dragged around the arrastra, reducing the ore to a fine mud. This was then spread over a courtyard or patio, sprinkled with mercury, salt, and copper sulfate, and mixed by repeatedly driving mules over it. Chemical reactions freed the silver from its compounds and caused it to dissolve in the mercury. When the amalgamation was complete, the material was agitated with water in large tubs and the mud run off. The amalgam remaining at the bottom was collected and heated to drive off the mercury. The process, especially suitable for the silver ores of the dry, barren areas of Mexico, was responsible for a large proportion of the world's silver production for 350 years; it was finally displaced by the cyanide process early in the 20th century.

Development of the Patio Process

Bartolomé de Medina was a successful Spanish merchant who became fascinated with the problem of decreasing silver yields from ores mined in Spanish America. By the mid-sixteenth century, it was well known in Spain that American silver production was in decline due to the depletion of high-grade ores and increasing production costs. The New Laws, prohibiting the enslavement of Indians, had resulted in higher labor costs as miners turned to wage labor and expensive African slaves. These higher production costs made mining and smelting anything but the highest grade silver ores prohibitively expensive, just as the availability of high grade ores was in decline. Bartolomé de Medina initially focused his attentions on learning about new smelting methods from smelters in Spain. He was approached during his research by an unknown German man, known only as "Maestro Lorenzo," who told him that silver could be extracted from ground ores using mercury and a salt-water brine. With this knowledge, Medina left Spain for New Spain (Mexico) in 1554 and established a model patio refinery in order to prove the effectiveness of the new technology. Medina is generally credited with adding "magistral" (a copper sulfate) to the mercury and salt-water solution in order to catalyze the amalgamation reaction. However, some historians assert that there were already sufficient copper sulfates in the local ores and that no additional magistral was needed. Regardless of whether or not Medina's contribution was entirely original, he promoted his process to local miners and was able to obtain a patent from the Viceroy of New Spain. As a result, he is generally credited with the invention of silver amalgamation in the form of the patio process.

Basic Elements of the Patio Process

Patio Process at the Gould & Curry Mill, Comstock Lode, USA, 1866

Silver ores were crushed (typically either in "arrastras" or stamp mills) to a fine slime which was mixed with salt, water, magistral (essentially an impure form of copper sulfate), and mercury, and spread in a 1-to-2-foot-thick (0.30 to 0.61 m) layer in a patio, (a shallow-walled, open enclosure). Horses were driven around on the patio to further mix the ingredients. After weeks of mixing and soaking in the sun, a complex reaction converted the silver to native metal, which formed an amalgam with the mercury and was recovered. The amount of salt and copper sulfate varied from one-quarter to ten pounds of one or the other, or both, per ton of ore treated. The decision of how much of each ingredient to add, how much mixing was needed, and when to halt the process depended on the skill of an *azoguero* (English: quicksilver man). The loss of mercury in amalgamation processes is generally one to two times the weight of silver recovered.

The patio process was the first form of amalgamation. However, it is unclear whether this process or a similar process—in which amalgamation occurred in heated vats rather than open patios—was the predominant form of amalgamation in New Spain, as the earliest known illustration of the patio process dates from 1761. There is substantial evidence that both processes were used from an early date in New Spain, while open patios were never adopted in Peru. Both processes required that ore be crushed and refiners quickly established mills to process ore once amalgamation was introduced. By the seventeenth century, water-powered mills became dominant in both New Spain and Peru.

Due to amalgamation's reliance upon mercury, an expansion of mercury production was central to the expansion of silver production. From shortly after the invention of mercury amalgamation to the end of the colonial period, the Spanish crown maintained a monopoly on mercury production and distribution, ensuring a steady supply of royal income. Fluctuations in mercury prices generally resulted in corresponding increases and decreases in silver production.

Broader Historical Significance

The introduction of amalgamation to silver refining in the Americas not only ended the mid-sixteenth century crisis in silver production, it also inaugurated a rapid expansion of silver production in New Spain and Peru as miners could now profitably mine lower-grade ores. As a result of this expansion, the Americas became the primary producer of the world's silver, with Spanish America producing three-fifths of the world's silver supply prior to 1900.

While a number of factors resulted in the minimal use of forced Indian labor in Mexican silver production, the introduction of silver amalgamation allowed for an expansion of silver production in Peru that had profound consequences for Peru's native population. Francisco Toledo, the Viceroy of Peru in the 1570s, saw amalgamation as the key to expanding Peruvian silver production. He encouraged miners to adopt amalgamation and construct mercury mines. More significantly, in order to provide sufficient labor to accommodate the expansion of silver mining to lower-grade ores, Toledo organized an Indian draft labor system, the mita. Under this system, thousands of natives were forced to work in silver and mercury mines for less than subsistence-level wages. Twelve thousand draft laborers regularly worked at the largest mine in the Americas, located at Potosí in modern Bolivia. Native attempts to avoid the mita led to the abandonment of many Indian villages throughout Peru. Spanish monopolization of refining through amalgamation cut natives out of what had earlier been a native-dominated enterprise. Refining represented the most profitable segment of silver production. In conjunction with the mita, the exclusion of natives from owning refineries contributed to the transformation of Peruvian natives into a poorly paid labor force.

The rapid expansion of silver production and coinage—made possible due to the invention of amalgamation—has often been identified as the primary driver of the price revolution, a period of high inflation lasting from the sixteenth to early seventeenth-century in Europe. Proponents of this theory argue that Spain's reliance on silver coins from the Americas to finance its large balance of payments deficits resulted in a general expansion of the European money supply and corresponding inflation. Critics of the theory, however, argue that inflation was really a result of European government policies and population growth.

While the role of the expansion in silver production in the price revolution may be disputed, this expansion is often acknowledged as a key ingredient in the formation of early-modern world trade. Spanish American production fed Chinese demand for silver, facilitating the development of extensive trade networks linking Europe, Africa, Asia, and the Americas as Europeans sought to gain access to Chinese wares.

Pidgeon Process

The process involves reducing molten magnesium oxide slag by ferrosilicon under low gas pressure at a temperature of about 1400°C.

The metallic magnesium, formed in the process, evaporates and then condensates away from the hot region.

The condensed magnesium, having purity of 99.95% is then remelt and cast.

The Three Steps of the Pidgeon Process:

1. Calcination of Dolomite, at 1300° C, Highly Endothermic Reaction, CO2 Released both from the reaction, and from Fuel Burned for Process Heat.

   $CaMg(CO_3)_2 \rightarrow CaO + MgO + 2CO_2$

2. Ferrosilicon Alloy Production, the Ferrosilicon Alloy Production by Electric Arc through Mixture of Hematite, Quartz Sand, and Coal. Extremely Endothermic Reaction; Emits Toxic CO.

   $Fe_2O_3 + 4SiO_2 + 11C \rightarrow 2(Fe)Si_2 + 11CO$

3. Silicothermic Reduction of MgO by Ferrosilicon, at ~14000 C under Vacuum for 8- 10 hrs. Highly Endothermic Reaction, the ratio of feed are 6 calcine dolomite: 1 ferrosilicon as briquettes.

   $2MgO + 2CaO + (Fe)Si \rightarrow 2Mg(g) + Ca_2SiO_4(s) + Fe$

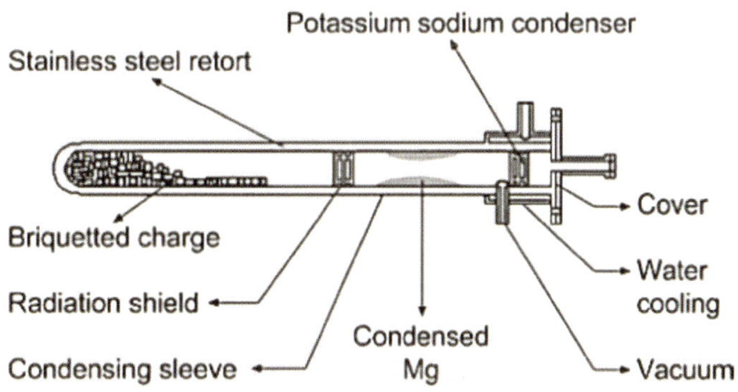

Hoopes Process

Hoopes process is mainly used on primary aluminium to produce extremely pure aluminium. This process is structured so that it consists of three liquid layers, which are separated due to different densities of the layers. This is illustrated in figure below. The bottom layer consists of an aluminium copper alloy mixed with the impure aluminium, and serves as the anode. The scrap is mixed with the copper to make it denser. The top layer consist of pure aluminium and serves as the cathode. These two layers are separated by a layer of molten electrolyte. The mid layer can consist of different substances, but normally this is a layer of molten inorganic salts, usually a chloride or fluoride. The idea of this process is to supply sufficient energy for the aluminium to react with the electrolyte and thus be transferred from the bottom layer up to the middle layer. When this happens, all elements more reactive than aluminium will transfer as well. Then the electrolyte is composed of components which form more stable compounds with the elements more reactive than aluminium, so only pure aluminium is able to travel up to the top layer. This way pure Al^{3+} is transported all the way to the top layer and can be removed as very pure aluminium.

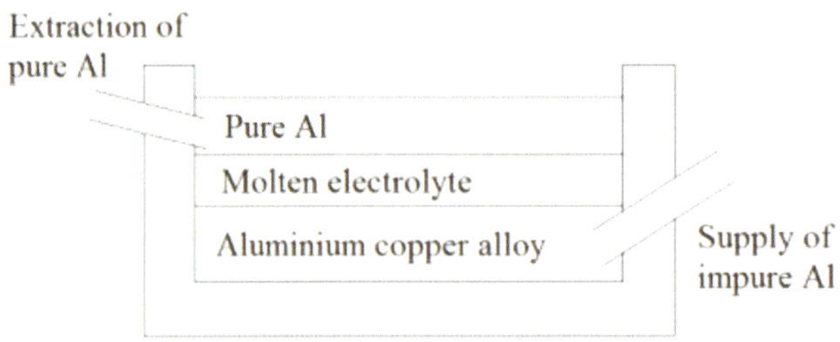

Figure: Principles of three-layered electrolysis.

The temperature requirement for this process is 700-900°C, and the energy use is between 17- 18 kWh/kg. The process can produce aluminium with a purity of >99.07 %. It is assumed that this energy requirement include both energy to reach desired temperature, to maintain this temperature throughout the process and the actual energy consumed by the process.

Schwartz and Wendt states that this process is able to achieve a purity level comparable to primary aluminium even for scrap with a purity level as low as 70 %.

The electrolyte and copper alloy used for this process can be reused and is thus not consumed during this process. They do however require some processing to get rid of the trapped impurities before they can be reused.

Adjusted Three-layered Electrolysis

As mentioned the three-layered electrolysis is mainly applied to primary aluminium to produce extremely pure aluminium. Kjos conducted an experiment to see if the three-layered electrolysis could be adjusted to fit aluminium scrap and possible reduce the energy requirements if the purity demands of the refined aluminium were lowered. They tested redesign of the cathode to reduce losses through voltage drop, and a new method redesigning the setup of the electrolysis. This is from a confidential report, and therefore only the energy use achieved can be included here. An energy use of respectively 8.5 and 5 kWh/kg were achieved.

Dow Process

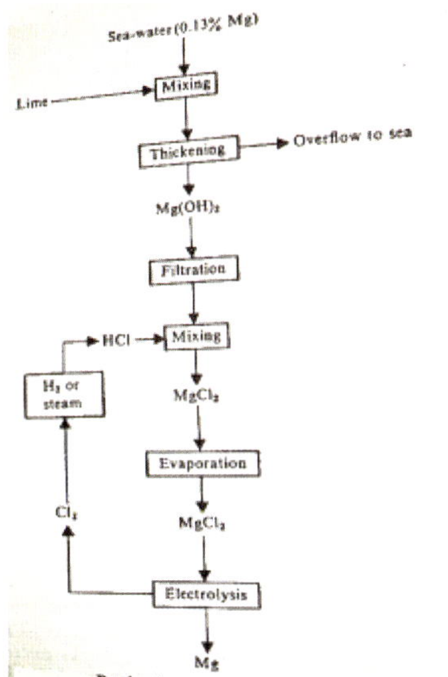

Production of Magnesium from Sea-Water.

The Dow Process is generally applied for extraction of Magnesium from Sea water. In this process first we add lime for thickening then mix with 10% HCl.

The electrolytic cell consists of a brick-lined vessel, divided into anode and cathode compartments by a semi-wall. Air or watercooled Graphite plate anode and steel cathode are submerged in electrolyte composed of alkaline chlorides with addition of magnesium chloride. The operating temperature is 680°C to 750°C. Magnesium chloride decomposes in the electrolytic cell according to the reaction

$$MgCl_2 \rightarrow Mg + Cl_2\,(g)$$

Metallic magnesium is formed at the cathode. It floats up (it is lighter than electrolyte) collecting in the cathode compartment. Chlorine, which is by-product of the process, is collected in the anode compartment.

In this method CaCl2 and NaCl are added to anhydrous MgCl2 to decrease its melting point.

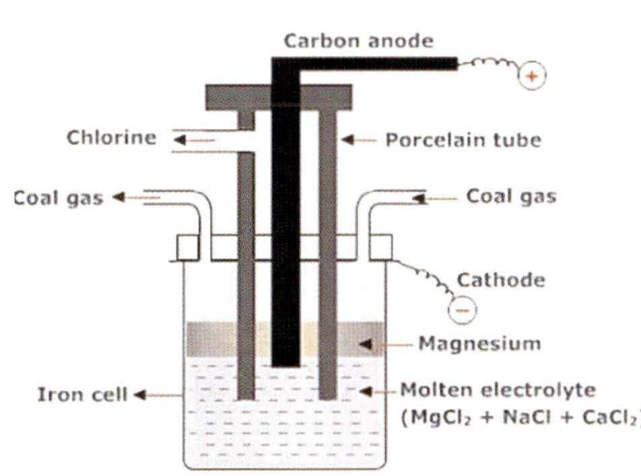

Dow Cell

References

- Douglas Fisher, "The Price Revolution: A Monetary Interpretation," The Journal of Economic History, Vol. 49, No. 4 (Dec. 1989), 883-884.
- The-Bessemer-Process: wirralmodelengineeringsociety.co.uk, Retrieved 16 April 2018
- The-refining-of-lead-by-the-betts-process: slideheaven.com, Retrieved 12 May 2018
- Dennis Flynn and Arturo Giráldez, "Born with a 'Silver Spoon': The Origin of World Trade in 1571," Journal of World History, Vol. 6, No. 2 (Fall, 1995), 202-203.
- Patio-process, technology: britannica.com, Retrieved 30 March 2018

Permissions

All chapters in this book are published with permission under the Creative Commons Attribution Share Alike License or equivalent. Every chapter published in this book has been scrutinized by our experts. Their significance has been extensively debated. The topics covered herein carry significant information for a comprehensive understanding. They may even be implemented as practical applications or may be referred to as a beginning point for further studies.

We would like to thank the editorial team for lending their expertise to make the book truly unique. They have played a crucial role in the development of this book. Without their invaluable contributions this book wouldn't have been possible. They have made vital efforts to compile up to date information on the varied aspects of this subject to make this book a valuable addition to the collection of many professionals and students.

This book was conceptualized with the vision of imparting up-to-date and integrated information in this field. To ensure the same, a matchless editorial board was set up. Every individual on the board went through rigorous rounds of assessment to prove their worth. After which they invested a large part of their time researching and compiling the most relevant data for our readers.

The editorial board has been involved in producing this book since its inception. They have spent rigorous hours researching and exploring the diverse topics which have resulted in the successful publishing of this book. They have passed on their knowledge of decades through this book. To expedite this challenging task, the publisher supported the team at every step. A small team of assistant editors was also appointed to further simplify the editing procedure and attain best results for the readers.

Apart from the editorial board, the designing team has also invested a significant amount of their time in understanding the subject and creating the most relevant covers. They scrutinized every image to scout for the most suitable representation of the subject and create an appropriate cover for the book.

The publishing team has been an ardent support to the editorial, designing and production team. Their endless efforts to recruit the best for this project, has resulted in the accomplishment of this book. They are a veteran in the field of academics and their pool of knowledge is as vast as their experience in printing. Their expertise and guidance has proved useful at every step. Their uncompromising quality standards have made this book an exceptional effort. Their encouragement from time to time has been an inspiration for everyone.

The publisher and the editorial board hope that this book will prove to be a valuable piece of knowledge for students, practitioners and scholars across the globe.

Index

P

R

S

T

Z